out of print   x 10-

8W
141V 1975

Science Since Babylon

Science Since Babylon

Enlarged Edition

Derek de Solla Price

New Haven and London

Yale University Press

1975

Printed in the United States of America by
The Colonial Press, Inc., Clinton, Massachusetts

Published in Great Britain, Europe, and Africa by
Yale University Press, Ltd., London.
Distributed in Latin America by Kaiman & Polon,
Inc., New York City; in Australasia and Southeast
Asia by John Wiley & Sons Australasia Pty. Ltd.,
Sydney; in India by UBS Publishers' Distributors Pvt.,
Ltd., Delhi; in Japan by John Weatherhill, Inc., Tokyo.

Permission to reprint "Automata and the Origins
of Mechanism and Mechanistic Philosophy,"
© 1964 by the University of Chicago Press,
"Geometrical and Scientific Talismans and
Symbolisms," © 1973 by Heinemann Educational
Books Ltd., and "The Difference Between Science and
Technology," first published by the Thomas
Alva Edison Foundation, is gratefully acknowledged.

To the memories of

JOHN FULTON
and
CHARLES SINGER

beloved pioneers in the history
of science and medicine

Contents

Preface to Enlarged Edition

PERHAPS just because it was not designed as a comprehensive and complete history of science, or even of any period or part of science, this book seems to have survived and still be in demand to serve its purpose as a set of examples. Though the intervening fourteen years have seen huge and irreversible changes in the public mind concerning science, all the doubts and disaffection have only sharpened the urgent need for understanding and perceptive analysis to replace the mess of convenient superstitions surrounding the relations between science and society.

History of science as a professional field has flourished reasonably well, and Yale's department with it, though we still wince a little if enquirers express surprise about it being a separate department and a field of knowledge not contained in history, not wedded to philosophy. In other places such dualities certainly occur, but our own faculty and a decade's-worth of our Ph.D.'s occupy a wide range of honorable niches and share only the qualification that they are professionals in the history of science, medicine, and technology. For my own part, as the field has grown I have become increasingly conscious and increasingly active in trying to meet the challenge of modern problems in science policy. After all, in a sense, historical understanding may be looked on as an attempt to predict the past, and if that can be done, the same basis of analysis may be used to make reasonable second guesses about the afflictions of the present.

If I had the job to do again, these lectures might insist a little less on the humanities of science and a little more on our fields as vital to science policy studies. Today as never before, our higher educational system and the culture it enfolds teeter critically on a sharp division between education in the ancient sense of the term and a somewhat blatantly utilitarian viewpoint in which science is seen as a begetter of technological fixes for national needs. Curiously enough, the most leftist and most rightist commentators coincide in this latter attribution, which I believe to be dangerously and misleadingly wrong. As I have tried to show in these following chapters, science is a very exceptional and peculiar activity of all mankind, and one is not at liberty to regard it as that which can be applied to make technology. There is not even any simple relationship between the two, though scientists often pretend there is to ease their dealings with politicians and administrators. The true relationship is complex. We still do not know the answers to many of its puzzles and there is today even more need for historical understanding and empirical studies to help solve these problems.

To report the continuing saga of some of the research topics here discussed, I have added postscripts to the original chapters and a few minor corrections and explanations. In addition, three new pieces for which I have had numerous requests have been added to the collection. A study of the history of automata serves as a link between the development of clockwork and the mechanistic philosophy that has played a central role in the conceptual side of science. Another study of geometrical amulets links science with magical pseudoscience and amplifies the first chapter, in which two different modes of scientific thought had been

explored historically. The third extra section deals specifically with the difficult matter of the relation between
science and technology.

I am grateful to the University of Chicago Press, the D.
Reidel Publishing Co., Inc. (Heinemann Educational
Books, Ltd.), and the Thomas Alva Edison Foundation,
for permission to reproduce corrected versions of these additional sections from their previously published forms.

D. de S. P.

New Haven, Connecticut
July 1, 1974

Preface to Original Edition

THIS BOOK had its origin in a set of five public lectures given at the Sterling Memorial Library at Yale University during October and November 1959 under the auspices of the Yale Department of History. At that time, they were designed not for publication but rather to attract the attention of humanists and scientists by oral presentation to what is usually called the History of Science, but for which I prefer the eccentric but broader term Humanities of Science.

The subject (whatever its name) had just come through a stage in its growing up during which it almost seemed as though every would-be practitioner of the art deemed it necessary to exhibit the completeness of his dedication by writing *the* history of the whole of science through all its periods. Hoping that this historiographic phase had evaporated, and feeling incompetent in too many scientific and historical directions, I resolved instead to essay the experiment of speaking only from those areas in which I had reasonable firsthand experience at research.

Within this limitation I strove to cover the gamut of the historical range from the Babylonians to the near future, bringing in as many fields of application as possible in the hope of showing humanists that our new discipline might make an interesting neighbor to their own. I hoped, in addition, to show scientists that we ought to be able to talk about science with as much scholarly right as other humanists receive, and that our approach might (if successful) lead to a different or better understanding than one could get by just "doing" science. To the educators I tried to show that this subject was the missing bridge that would allow the good liberal education to include some mention of science, and to do it with genuine scholarship instead of by watering down science for humanistic babes or dishing up Greek sculpture in the hope of rearing cultivated scientists.

Since I had deliberately restricted myself to my own research experience, which has been oddly varied, my topics are none of them well-worn paths within the history of science. It was with some dejection that I had to forgo the undeserved privilege of speaking as a proper medievalist, an expert on Newton, Galileo, or Darwin, a historian of alchemy or of mechanics or of any of the other thoroughbreds. To my colleagues I can only apologize if their field has thereby been misrepresented to the world at large. I should be sorry, because in that case they should lose my sympathy for allowing themselves to become too circumscribed by the traditions of a subject young enough to leap over such bounds.

This book, then, is no comprehensive history of science, or even a partial history of sundry fragments of science, its theories, or its personalities. Since the lectures had to have some thematic connective tissue, I took the egotistic liberty of considering that the things I had been interested in were all *crises:* that they were all somehow vital decisions that civilization had to make to turn out the way it did and lead us to our present scientific age. The first crisis, described in the initial lecture, had to be that which made our own civilization start to become scientific, thereby setting it apart from all other cultures. The second lecture deals with the departure of science from the realm of pure thought and its transformation into scientific technology. The third pursues this technological thread back into the web of Renaissance and modern science. The fourth pin-points the stark transition from classical theories in the nineteenth century to the explosive multiplication of discoveries of the twentieth. The last topic represents an attempt to draw upon history and science to make a guess about the probable transition from this present state to a future internal economy of science that already looks quite different.

The chief reasons for publishing such a set of lectures are, positively, that they were successful and, negatively, that they should not be made up into a longer book. On the credit side, the example and special pleading of the lecture series may have helped, and certainly did not ultimately hinder, John Fulton, Professor of the History of Medicine, in his great ambition of achieving at Yale a full and autonomous Department of the History of Science and Medicine —a department in which I rejoice to have my present post. On the other side of the ledger, because of my manner of choosing topics it seems plain that each lecture should eventually become a separate monograph, relieved of the obligation of unity when placed side by side with the others. Expansion to greater dimensions at this stage therefore seemed purposeless. I have instead added an Epilogue describing this fruition of the lectures: a teaching and research department that will use all the humanistic techniques in an analysis of science. Like the rest of the book, it is a very personal testament. Unlike the rest, it is based not on any research experience, conventional or unconventional, but upon hope—and on a strong conviction—that science is so important in our lives that all weapons in our battery of criticism of it must be well manned. It is not sufficient that historians of science must exist (though that seems difficult enough); but they must soon take their place at the forefront of scholarship to help preserve and advance all that we hold dear in civilization.

Among all the debts I have to pay, I should like to record the following: To the high school teachers that had the wisdom to give me de Moivre before brackets and Anglo-Saxon before Shakespeare. To Cyril Parkinson, in Malaya, who conflated badminton, history, and exponential growth. To Dr. Harry Lowery, who gave me the best practical education as a physicist. To Christ's College, Cambridge, and to Sir Lawrence Bragg, whose kindness and hospitality meant so much in the Cavendish Laboratory. To Robert Oppenheimer and the Institute for Advanced Study for the opportunity to work there on a Donaldson Fellowship for two glorious years, and to Otto Neugebauer for guiding me there. To the Commonwealth Fund Fellowship that first brought me to America, and to the Smithsonian Institution that brought me here again. To David Klein, whose delectable knowledge of the power of words intervened to save this book from so many breaches of syntax and good taste. To all my new friends at Yale who kept me here and have given me so much to be excited about.

D. de S.P.

New Haven, Conn.
September 15, 1960

CHAPTER 1

The Peculiarity of a Scientific Civilization

WHEN that prodigious oddity of an Indian mathematician Srinivasa Ramanujan lay mortally ill at the age of thirty in a London hospital, he was visited by one of his peers, Professor G. H. Hardy of Cambridge. Wishing to divert the patient, and being at that time absorbed with number theory, Hardy remarked that he had just driven up in a taxicab numbered 1729 and that the number seemed to be rather a dull one. "Oh, no!" replied Ramanujan. "It is a very interesting number; it is actually the smallest number expressible as a sum of two cubes in two different ways." [1]

One might suppose that this story,[2] like the number itself, is just a trivial item in the anecdotage of great mathematicians. Oh no! It is actually a most nontrivial and indicative pathological example which may elucidate the highly gen-

1. To relieve any tension, let me volunteer the information that $1729 = 9^3 + 10^3 = 1^3 + 12^3$.

2. The biographical material on this most interesting of all mathematical prodigies has been reprinted in *Collected Papers of Srinivasa Ramanujan*, eds. G. H. Hardy, P. V. Seshu Aiyar, and B. M. Wilson (Cambridge, 1927), and in G. H. Hardy, *Ramanujan—Twelve Lectures Suggested by His Life and Work* (Cambridge, 1940). See also the article by James R. Newman in *The World of Mathematics*, 1 (New York, 1956), p. 368.

eral and fundamental problem of our scientific civilization.
We may call this problem that of accounting for the "peculiarity" of our modern culture, implying by the ambiguous
term not only that it is different from others but also that
it contains a novel and even bizarre element which distinguishes it from all that has gone before.

Thanks to the scholarship of the historian and archaeologist, we have today, pinned to the academic dissecting board,
a whole series of high civilizations about which we are beginning to know a great deal more than our forebears. We
have the Assyrians and the Egyptians, the Greeks and the
Romans, the Aztecs and the Incas, the Chinese and the Indians, the empire of Islam, and our own contemporary
world. Like George Orwell's animals, all these civilizations
are related, but some are more related than others. The
ones about which we know most are those that lead in direct chronological sequence through the stages of Greece,
Rome, Byzantium, and Islam to our Middle Ages, Renaissance, Industrial Revolution, and present culture. Each
of these has enough characteristics and peculiarities for us
to label it as an entity for historical expediency, but it is
also clear that together they stand as a related family, in-
heriting from generation to generation. Apart from this
family are a few great civilizations that seem each to stand
in relatively greater isolation. These have been exposed to
clear and detailed view only in quite recent times.

I suggest that our new-found knowledge of these more
isolated civilizations makes it meaningful to ask a question
that could neither arise nor be answered earlier, when we
were conscious only of the Greek Miracle and our descent
therefrom. What is the origin of the peculiarly scientific
basis of our own high civilization? In our present generation
we may stand on the shoulders of giants and examine in
considerable detail the history of science in China, the complexities of Babylonian mathematics and astronomy, the

machinations of the keepers of the Mayan calendar, and the scientific fumblings of the ancient Egyptians. Now that we have some feeling for what was possible (and what not) for these peoples, we can see clearly that Western culture must somewhere have taken a different turn that made the scientific tradition much more productive than in all these other cases. We are now living in a high scientific technology, in which the material repercussions of science shape our daily lives and the destinies of nations and in which the philosophical implication of the Scientific Revolution, to quote Herbert Butterfield, "outshines everything since the rise of Christianity and reduces the Renaissance and Reformation to the rank of mere episodes, mere internal displacements, within the system of medieval Christendom." [3]

We know now that none of the other great civilizations followed a comparable scientific path. It becomes ever clearer from our fragmentary historical understanding of their case histories that none of them was even approaching it. Two distinct attitudes are possible. The conventional one is to examine each civilization in turn and to show how

3. Herbert Butterfield, *The Origins of Modern Science* (London, 1949), p. viii. The same author, most distinguished among living traditional historians, has more to say about the study of the history of science: "One of the safest speculations that we could make today about any branch of scholarship is the judgment that very soon the history of science is going to acquire an importance amongst us incommensurate with anything that it has hitherto possessed. It has become something more than a hobby for the ex-scientist or a harmless occupation for a crank; it is no longer merely an account of one of the many human activities like the history of music or the history of cricket—activities which seem to belong rather on the margin of general history. Because it deals with one of the main constituents of the modern world and the modern mind, we cannot construct a respectable history of Europe or a tolerable survey of western civilization without it. It is going to be as important to us for the understanding of ourselves as Graeco-Roman antiquity was for Europe during a period of over a thousand years." Quoted from the first Horblit Lecture on the History of Science, *Harvard Library Bulletin*, XIII (1959), 330.

the exigencies of wars and invasions, political and social conditions, economic disadvantage, or philosophic strait-jackets prevented the rise of any sort of Scientific Revolution. Perhaps it is a natural vanity to attempt to show that ours is the only one in step. A more rational alternative is to entertain the possibility that it is our civilization which might be out of step. Conceivably, the others were, for the most part, normal, and only our own heritage contained some intruding element, rare and peculiar, which mush-roomed into the activity that now dominates our lives. One may legitimately speculate about the rarity of science in civilizations, just as the astronomer may speculate about the rarity of planetary systems among suns, or the biologist about the rarity of life on planets.

Fortunately, we can do more than speculate if we under-stand something of the evolutionary mechanism, be it of planetary systems or living matter or scientific activity. Thus armed, we may make reasonable guesses as to what to seek by way of an origin of the phenomenon. To understand the place of science in our world today, then, we must trace back through the continuum of its history, seizing on the pivotal moments. These are not necessarily instances of great discoveries or advances; rather they are junctures at which men had to put on a new sort of thinking cap or in-ject some quite new element into their deliberations.

It is now quite reasonably established and agreed that modern science has developed in an orderly and regular fashion from the heyday of the Scientific Revolution (cen-tered in the seventeenth century) until the present day. Pivotal points there have certainly been, and we must later discuss some of them, but it seems as though a recognizable embryo of modern science was already present in the work of Newton. If it was there, it was also there earlier in that of Galileo and Copernicus and, to choose other fields, in that of Harvey and Boyle. The formal network of inter-penetrating theories, experiments, and concepts retained by modern science certainly includes these names, as any science teacher well knows. Indeed, to many teachers the history of science exists pre-eminently as a device for enhancing with a little human interest the names occurring in their pedagogic practice.

If, however, the embryo of modern science was already present in the sixteenth century, we must seek still earlier for the singular events attending its conception. What sort of events do we examine? Without some strategy of attack we become mere chroniclers and annalists of the several autonomous fields into which science is now divided. It is a delicately subtle historical error to carry back too rigorously the compartmentalization of science before the sixteenth century, when learning was much more a single realm and even the genius was a polymath.

It would be poor tactics in scholarship to attempt to span the whole wide front of knowledge, and some limitation is essential to attain perspective. Of all limited areas, by far the most highly developed, most recognizably modern, yet most continuous province of scientific learning, was mathematical astronomy. This is the mainstream that leads through the work of Galileo and Kepler, through the gravitation theory of Newton, directly to the labors of Einstein and all mathematical physicists past and present. In comparison, all other parts of modern science appear derivative or subsequent: either they drew their inspiration directly from the successful sufficiency of mathematical and logical explanation for astronomy or they developed later, probably as a result of such inspiration in adjacent subjects.

Here we must make a digression to exclude from this analysis a certain hard core of technology and science which every civilization must have and does usually attain as part of its necessary equipment. Men must always build shelters, raise crops and distribute them, break each others' heads, mend broken heads, and know why all these things should be done. Consequently, permeating all recorded history and all cultures, we find some knowledge of the basic geometry of houses and fields, of merchants' reckoning and calendar computation, of industrial chemistry and medical practice, and of a cosmology closely associated with religion. Each of these components of sciences is capable of being developed to considerable sophistication without resulting or even participating in a scientific revolution. As evidence may be cited the Mayan calendar, a maze of arithmetical juggling which permeated an entire culture without making it "scientific." Even the high arts of medicine and chemistry, which were already flourishing during the first few centuries before our era and grew steadily for nearly two millennia thereafter, did not change radically or begin to assume their modern scientific garb until they had been preceded by the revolution of the exact sciences.

Thus, for strategic reasons, we must fix our attention upon that one highly technical and recondite department of science which served as a matrix for the theories of Copernicus and of Kepler and provided the raw material for the first extraordinary conquests of the Scientific Revolution, standing as a model and encouragement to the rest. Concentration on the pre-Copernican state of this mathematical astronomy carries us back at a single swoop to the Hellenistic period and has the additional advantage of providing the strongest link between ancient and modern science. This is the astronomical theory developed more fully in the *Almagest,* composed by Claudius Ptolemaeus (Ptolemy) about A.D. 140.[4]

The *Almagest* is second only to Euclid in its dominance

4. The *Almagest* is available in respectable editions only in the original Greek and in a German translation by Karl Manitius, Teubner Classics (1912), long out of print. The available editions in French, by the Abbé Halma (Paris, 1816), and in English in the *Encyclopaedia Britannica, Great Books of the Western World, 16* (Chicago, 1952) are full of errors, difficult of language, and a grave injustice to the most important book of science of the ancient and medieval world.

through the centuries. To the modern mathematician or scientist re-examining the technical content of the texts, both exhibit a depth of sophistication that re-emerged only quite recently. Euclid, being pure mathematics, is in a sense still with us today, albeit slightly battered by non-Euclidean geometers. The *Almagest*, being science, has been outmoded and lost to all but the historians of astronomy. Because of this, the ever-ready and popular mythology of science has attributed to Ptolemaic astronomy several features which are wrong or misleading. Wishful thinking, oversimplification, and the copying of secondary sources unto the *n*th generation are particularly rife when even scientists talk in an amateur way about science.

To clear the air we must remark that the main burden of the *Almagest* is to provide a mathematical treatment of the extremely complex way in which each of the planets appears to move across the background of the fixed stars. Relative to its times, the *Almagest* must have seemed as formidable and as specialized as Einstein's papers on relativity do to us. Both Ptolemy and Einstein have had their popularizers. The statements that "Einstein proved that everything is relative" and that "Ptolemy proved that everything rotates around the fixed earth" are equally inadequate irrelevancies. In point of fact, despite the obvious importance of his philosophical innovations in cosmology, Copernicus necessarily left the mathematical machinery of the *Almagest* unchanged and intact in all its technical essentials. Moreover, each small change he made slightly worsened the correspondence between theory and observation. Only for the motion of the moon (which is geocentric anyhow) was his theory superior if not original.[5]

5. For a more detailed description of the much misunderstood scientific status of the old and new planetary theories see Derek J. de S. Price, "Contra-Copernicus: A Critical Re-Estimation of the Mathematical Planetary Theory of Ptolemy, Copernicus and Kepler," in *Critical Problems in the History of Science*, ed. Marshall Clagett (Madison, 1959), pp. 197–218.

It is, then, in the *Almagest* that we see the triumph of
a piece of mathematical explanation of nature, achieved
already in the Hellenistic period and working perfectly
within the limits of all observations possible with the naked
eye. It was clearly the first portion of complicated science to
acquire a sensible and impressive perfection. Mathematical
planetary theory became very early in our history the one
region of knowledge of the physical world where the indis-
putable logic of mathematics has been proved adequate and
sufficient. It is the only branch of the sciences that survived
virtually intact when the Roman Empire collapsed and
Greek higher mathematics was largely lost. It retained its
power and validity even after Copernicus, being superseded
only by the more recondite mathematics of Kepler and the
splendidly direct visual proof lent by Galileo's telescope
after 1600. Even to the layman this queer subject of the
mathematics of planetary motion has been regarded through
the ages as a bright jewel of the human intellect, fascinating
people with the godlike ability of mortals to comprehend
a theory so bristling with abstruse complexities yet so demonstrably and certainly true.

It is reasonable, therefore, to hazard the guess that this
hard central theory constitutes an intellectual plateau in
our culture—a high plateau present in our civilization but
not in any of the others. In all the branches of science in
all the other cultures there is nothing to match this early
arrival of a refined and advanced corpus of entirely mathematical explanation of nature. If we have put our finger
on an oddity in our intellectual history, there is, however,
no guarantee that this is the local oddity that has given us
modern science. Is this any more than just a caprice of circumstance attending the development of one particular
science?

The answer must be sought by carrying back the analysis

to still earlier times. If the *Almagest* is seen to develop by steady growth and accretion, spiced with flashes of inspiration, the history is similar to that proceeding from Newton to Einstein and is reasonably normal. If, on the other hand, we can show the presence of some intrinsic peculiarity, some grand pivotal point, we may be sure that this is the keystone of our argument.

Until but a few decades ago there was not a glimmer to indicate that the Greek Miracle was anything other than the rather local and well-integrated affair that generations of study of the classics would have us believe. In the field of astronomy, in particular, it was reasonable and evident that understanding and mathematical handling of the phenomena had evolved gradually, from almost primitive, simple beginnings up to the culmination of the *Almagest* and its later commentators. Certainly there was a sufficiency of known names of mathematicians and astronomers who must have achieved something before Ptolemy, and there existed a great corpus of stories, some no doubt partly true, telling what these men were supposed to have discovered or done.

The beautiful feeling of close approximation to perfect knowledge was, however, tempered by the more cautious; they were a little regretful only of the short-changing of the historian by that peculiarly scientific phenomenon which allows one successful textbook to extinguish automatically and (in those times) eradicate nearly all traces of what had gone before. Thus, although the very success of Ptolemy meant that we could know only fragments of pre-Ptolemaic astronomy, there were good reasons for hoping that our ignorance hid nothing vital.

This hope was perturbed and now lies shattered by the discovery, since 1881, of a great corpus of Babylonian mathematics and astronomy, evidenced by numerous tablets of clay inscribed with cuneiform writings and extending in date from the Old Babylonian period of the second millennium B.C. to the Seleucid period in Hellenistic times.[6]

It suffices for our present purposes to note that Babylonian astronomy, especially in its Seleucid culmination during the last two or three centuries B.C., represents a level of mathematical attainment matched only by the Hellenistic Greeks, but vastly different in content and mode of operation. At the kernel of all Babylonian mathematics and astronomy there was a tremendous facility with calculations involving long numbers and arduous operations to that point of tedium which sends any modern scientist scuttling for his slide rule and computing machine. Indeed some of the clay tablets, presumably intended for educational purposes, contain texts with problems which are the genotypes of those horrors of old-fashioned childhood—the examples about the leaky baths being filled by a multiplicity of variously running taps, and the algebraic perversions (though here expressed more verbally than symbolically) with a series of brackets contained within more brackets *ad nauseam.*

That is admittedly the dull side of Babylonian mathematics. Its bright side was a feeling for the properties of numbers and the ways in which one could operate with them. One gets the impression that the manner in which Ramanujan worked—in perceiving almost instinctively the properties of numbers far from elementary and in having every positive integer as one of his personal friends—was

6. The reader is referred to the discussion of this field, by its greatest exponent, Otto Neugebauer, in *The Exact Sciences in Antiquity,* 2d ed. (Providence, 1957). Here he will find ample reference to source material on Babylonian mathematical sophistication, as well as such entertainingly cryptic snippets as the story (p. 103) of how the first astronomical tablets were deciphered by Fathers J. Epping (of Quito, Ecuador) and J. N. Strassmeier (of London) and published in (of all places) the Catholic theological periodical *Stimmen aus Maria Laach,* starting in 1881. A good account of Greek and Babylonian material is Asger Aaboe, *Episodes from the Early History of Mathematics* (New York: Random House, 1964).

the normal mode for a Babylonian. I do not wish to exaggerate more than is necessary for effect, or to imply that the ancients all had the genius of Ramanujan. It is plain, though, that their forte was in matters arithmetical, and in this they were supreme.

The origins of this facet of Babylonian civilization are hard to determine. Perhaps it was some peculiar national characteristic; perhaps some facility given by the accident of their writing in clay with little, uniform, countable, cuneiform wedges. Possibly there was some urgency in their way of life that required the recording and manipulation of numbers in commerce or religion. Such tenuous speculation seems not only dangerous but unprofitable in view of the fact that the Babylonians were by no means unique in this quirk of mind: the Mayan passion for numbers in their calendrical cycles is not far different, though nothing that we thus far comprehend quite matches the superb Babylonian sophistication.

For our purposes, what is significant about the Babylonian attitude toward astronomy is not any accident of its origin, but rather that it existed as a highly developed and penetrating *arithmetical* way of dealing with the motions of the sun and moon and planets. The Babylonians operated with the vital technique of a place value system for all numbers, integral and fractional. They made use of the very convenient sexagesimal base of sixty, which we still retain from their tradition in our angle measurement of degrees, minutes, and seconds and in our subdivision of the hour. Above all, they were able to make astronomical calculations without recourse (so far as we know) to any sort of geometrical picture or model diagram. Perhaps the nearest thing to their methods in modern mathematics is in Fourier analysis of wave motions, but here the mathematician thinks in terms of concepts of sine waves rather than the mere numerical sequence of the Babylonians.

It is inevitable that we should be drawn to compare the high science of the Babylonians with that of the Greeks. For each we can perceive something like a reasonably continuous tradition until the last few centuries B.C., when both are concerned with the same very natural problem of the maddeningly near-regular motion of the planets. By that time each is standing ready with a mature and abstruse scheme full of technical refinements and containing, coordinated within the scheme, all the most relevant observations and considerations that had accrued through the centuries.

It is one of the greatest conjuring tricks of history that these two contemporary items of sophistication are as different from each other as chalk from cheese. Spectacularly, where one has deep knowledge, the other has deeper ignorance, so that they discuss precisely the same basic facts in manners so complementary that there is scarcely a meeting ground between them. For all the Babylonian prowess in computation, one discerns no element of that method of logical argument that characterizes the Greek Euclid. One might go further and accuse the Babylonians of being totally ignorant and incompetent with geometry (or, more generally, with all *Gestalt* matters), but here one must exercise due caution and allow them the modicum of architectural geometry, mythical cosmology, etc., that any high civilization seems to develop willy-nilly. For example, we know that the "Pythagorean" properties of the sides of the right-angled triangle were known to the Old Babylonians about a thousand years before Pythagoras; but this is precisely the sort of homespun geometry that can readily be acquired, even today, by anyone meditating upon a suitable mosaic floor or in a tiled bathroom.

What now of the Greeks? Are they not just as lopsided, scientifically speaking, as the Babylonians? We must be careful here to distinguish between the early Hellenic period and the later Hellenistic. In this distinction the history of science provides (as it does so often elsewhere) a perspective refreshingly different from that of other histories. For example, the great Renaissance, beloved of the historian of art, seems to diminish a little when viewed by the historian of science and to take much more of the character of a parochial Italian movement whose significance is overshadowed for us by the influential revival of astronomy in Protestant Germany.[7] As for the Greeks, the great centuries of art, philosophy, and literature of the Golden Hellenic age are overshadowed for us by the tremendous scientific vitality of the Hellenistic period.

Making what we can of the earlier period, we can discern the presence of an aura of logic and of geometry that we know so well from Euclid, but totally lacking is any depth of knowledge of calculation. Again, one must make the exception of the everyday and allow that an inhabitant of classical lands could, when pressed to it, function sufficiently to make out his laundry bill. One also allows the minute amount of arithmetic (in the Babylonian sense) contained in the well-known Pythagorean writings. Although these were concerned with number, and at times more than trivial, they were devoid of any difficult computation or any knowledge of the handling of general numbers far beyond ten. One need only examine the attitudes of each civilization toward the square root of two. The Greeks proved it was irrational; the Babylonians computed it to high accuracy.[8]

7. For a typical re-evaluation of the Renaissance as seen by the man who did more than any other to found the history of science as a scholarly autonomy, see George Sarton, *Appreciation of Ancient and Medieval Science during the Renaissance* (Philadelphia, 1955); especially the Epilogue, pp. 166–75.

8. For the Greek approach to the irrationality of the square root of two,

Again, we need here have little interest in understanding the series of complex motivations and accidents that had set the Greeks on this particular road of civilization. So far as it concerns science, other civilizations had probably done much this sort of thing before. Modern historians have long lived in consciousness only of the glorious and unique Greek tradition of mathematical argument from which we patently derive so much of our present state of mind; this being so, it is difficult to disabuse ourselves of the tradition and attempt to re-estimate how far Hellenism would have taken us in the absence of the Babylonian intervention so clearly manifest in such later Hellenistic writers as Hero and Hipparchus.

To cap the whole story, we now know that, to some extent at least, ancient Chinese civilization had grown up in effective isolation from both Babylonian and Greek, but with a steady development of arithmetical skills on the one hand and geometric on the other.[9] Is it not a mystery that, having both essential components of Hellenistic astronomy, they came nowhere near developing a mathematical synthesis, like the *Almagest,* that would have produced, in the

---

the most elegant short statement of the proof of Pythagoras is given by G. H. Hardy, *A Mathematician's Apology* (Cambridge, 1948), pp. 34–6; reprinted in James R. Newman, *The World of Mathematics, 4* (New York, 1956), 2031. The Babylonian approach is best seen in the tablet, Yale Babylonian Collection, No. 7289, and is commented upon by O. Neugebauer, *Exact Sciences in Antiquity* (Providence, 1957), p. 35 and pl. 6a. This tablet incidentally includes a geometrical diagram, though not as any aid to computation.

9. The story of Chinese mathematics, in all its branches, is now exhibited for the first time in Joseph Needham's *Science and Civilization in China, 3* (Cambridge University Press, 1959), section 19. The concluding chapter of this section (pp. 150–68), though it puts different emphasis on the place of political and philosophical conditions in East and West, comes to what is essentially the same conclusion I have reached here. Needham says, ". . . [in China] there came no vivifying demand [for mathematics] from the side of natural science. . . ." In the West, this demand arose through the strength of the mathematical planetary astronomy.

fulness of time, a Chinese Kepler, Chinese Newton, and Chinese Einstein? [10]

Let us look once more at the worlds of the Hellenic Greeks and the Seleucid Babylonians. It seems likely that they were in relatively little scientific contact before the great melting pot resulted from the unprecedented conquests of Alexander the Great, starting in 334 B.C. During succeeding centuries one discerns the entry into Greek mathematics and astronomy of results and methods so foreign and arithmetical that they could only have been lifted from Babylonian roots. Alas, apart from a few Greek writings in which "Chaldean" astronomers are cited in general or by name, we know little of the historical interaction and scientific marriage of these very different cultures. We can see only that it must have been supremely exciting to grapple with the end results of a science as alien to one's own as the Martians' but concerned with, and perhaps slightly more successful in treating, the same problems.

This is surely a spectacular accident of history that is powerful enough to stand as the pivotal point and provide thereby both proof and understanding of the essential peculiarity and difference of our own civilization from all others, even from the Chinese, which may have contained the same scientific elements but lacked the explosive im-

10. Mention of a Chinese Einstein prompts me to cite here the text of a letter by the Western Einstein, often quoted anecdotally but not, to my knowledge, ever given *in extenso*. I am grateful to my colleague, Professor Arthur Wright, for lending me a copy of the original, sent to Mr. J. E. Switzer of San Mateo, California. "Dear Sir, Development of Western Science is based on two great achievements, the invention of the formal logical system (in Euclidean geometry) by the Greek philosophers, and the discovery of the possibility to find out causal relationship by systematic experiment (Renaissance). In my opinion one has not to be astonished that the Chinese sages have not made these steps. The astonishing thing is that these discoveries were made at all. Sincerely yours /s/ A. Einstein. April 23, 1953."

pact between equal and opposite insights of Greek and Babylonian.

Our record of the accidents attending the birth of our own scientific civilization is incomplete at this time. Had it not been for further oddity, the stimulation of Greek geometry and logic by Babylonian numerical and quantitative methods might have been a mere flash in the pan, leaving behind nothing but a legacy of oriental schoolboy problems in the books of Hero of Alexandria and of Babylonian cycles of months in the calendar. The true fruition came as a natural but fortuitous consequence of combining the qualitative, pictorial models of Greek astronomical geometry with the quantitative operations and results of the Babylonians.

From the Greek point of view, the planets appeared to rotate almost, but not quite, uniformly in circles. By the Babylonians, the extent of the lack of uniformity was well measured and accurately predictable. How could the Greeks picture this slight but precise lack of uniformity in planetary motion? They could not conveniently do it by letting the planet move sometimes faster, sometimes slower. Motion that got faster and faster might be allowable, but there was no convenient mathematical machinery for considering a fluctuating speed. The most natural thing to do was to retain the perfect and obvious uniform motion in a circle and to let the Earth stand to one side of that circle, viewing the orbit with variable foreshortening. Such a theory accounts for the most complex actual motion, as we now know it, to an accuracy virtually as great as the eye can perceive without the aid of the telescope. Kepler showed that the planets move in an ellipse with the sun at one focus. The ellipse is, of course, very close to an off-center circle, and the planet appears to move with very nearly uniform angular velocity about the empty focus.

The Babylonian technique was to use series of sequences

composed of numbers that rose and fell steadily or had differences that themselves increased or decreased steadily. All the numerical constants were most cunningly contrived so as to yield the necessary periodicities and provide quantitatively accurate results without the intervention of any geometrical picture or model.

Thus the Greeks had a fine pictorial concept of the celestial motions, but only a rough-and-ready agreement with anything that might be measured quantitatively rather than noted qualitatively. The Babylonians had all the constants and the means of tying theory to detailed numerical observations, but they had no pictorial concept that would make their system more than a string of numbers.

This extraordinary mathematical accident of doing the only obvious thing to reconcile Greek and Babylonian, and deriving thereby a theory that was a convincing pictorial concept and also as near true as could be tested quantitatively, was a sort of bonus gift from nature to our civilization. As a result of that gift and its subsequent Hellenistic elaboration by trigonometrical techniques, the great book of the *Almagest* could stand for the first time as a complete and sufficient mathematical explanation of most complex phenomena. In nearly every detail it worked perfectly, and it exemplified an approach which, if carried to all other branches of science, would make the whole universe completely comprehensible to man. It stood also as a matrix for a great deal of embedded mathematical and scientific technique which was preserved and transmitted in this context up to the seventeenth century.

We must now survey our story and draw what conclusions we may. The fact that our civilization alone has a high scientific content is due basically to the mixture at an advanced level of two quite different scientific techniques—the one logical, geometrical, and pictorial, the other quantitative and numerical. In the combination of both approaches to astronomy, a perfect and workable theory was evolved, considerably more accurate than any other scientific theory of similar complexity. If one may speak of historical events as improbable, this Ptolemaic theory was improbably strong and improbably early. It was almost as though that branch of science had got an unfair start on all the others, racing ahead long before it should have in the well-tempered growth of any normal civilization, like the Chinese.

This interpretation should rather change the conventional attitude of historians toward the analysis of what happened in other regions of science. It has become usual to refer to the postponed scientific revolution in chemistry and the still more delayed freeing of the life sciences from their primitive states, and then to seek reasons for the tardiness of these changes. Once more this conventional attack may be fruitlessly seeking an explanation for what was, after all, the normal way of growth. Physics was forced early by the success of its neighbor subject astronomy, and when chemistry and biology develop, it seems very much as if the motivating forces are not internal but rather a pressure from the successes of physics and later chemistry.

Such an historical explanation, of course, begs the question of whether the priority of mathematical methods in astronomy was merely chronological or whether there exists also some fundamental way in which mathematical expression of the observed world is logically basic to our understanding, necessary whatever the historical accidents of growth. Philosophers of science usually consider only the latter possibility; science, as it is known to us, has an essential mathematical backbone. Since the historical origin of that backbone seems such a remarkable caprice of fate, one may wonder whether science would have been at all possible and, if so, what form it might have taken if (to make a hypothetical construct) a situation had existed in China which caused the chemical and biological sciences to make great advances before astronomy and physics.

If we are more satisfied and curious about the state of that science that we actually have, rather than what might have been, perhaps it behooves us to analyze further the consequences of our twin origin in the Graeco-Babylonian melting pot. It is more than a curiosity that of two great coeval cultures the one contained arithmetical geniuses who were geometrical dullards and the other had precisely opposite members. Are these perhaps biological extremes, like male and female, with comparatively little likelihood of an hermaphrodite? Possibly there is some special quality of nature or nurture that can make a human being, or even a whole society, excel in one of these extreme ways. Perhaps some men can excel in both, as Ptolemy evidently did. Perhaps the vigor of modern mathematical physics, for example, would demand that it be maintained by men who manage to excel both as Babylonians and as Greeks.[11]

Of some interest to the philosopher of science may be the suggestion from historical evidence that model-making in scientific theories and the use of quantitative methods may be a pair of complementary operations in the derivation of modern science. It is surely poetic justice that Niels

11. Shortly after writing these lines I happened to see a most sensitive paper by G. L. Huxley, "Two Newtonian Studies," in *Harvard Library Gazette, 13* (1959), 348–61. It ends with these words: "Last of the Babylonians indeed; but also the greatest of the Hellenic Geometers." It would indeed be rather interesting to determine if there was ever any other mathematician who did not betray himself as a lesser genius than usual when faced with either the Babylonian, analytical side of the subject or the Greek imagery of geometric thought and intuition. Huxley's comment is, of course, in part a reflection from that of Keynes: "Newton was not the first of the age of reason. He was the last of the magicians, the last of the Babylonians and Sumerians, the last great mind which looked out on the visible and intellectual world with the same eyes as those who began to build our intellectual inheritance rather less than 10,000 years ago." (*Royal Society Newton Tercentenary Celebrations,* Cambridge, 1947, pp. 27–34.)

Bohr's "Principle of Complementarity" holds sway so strongly just in that field of quantum mechanics where it is notorious that the visual aid of a physical model—a typically Greek device—has proved to be a snare and a delusion that must be banished from the scene.

The example of Ramanujan indicates that perhaps there are Babylonians, almost of pure mathematical breed, abroad among us today. Other mathematicians may surely be classed as of the Greek temperament. Unfortunately there has been very little useful study of the psychology of scientists, but the little that we have accords well with the notion that visual-image worshippers and number-magic prodigies may be surprisingly pure as strains.[12] Certainly we know from experience in the world of education that our population at large consists of those who take to mathematics and those who definitely do not. The problem is evidently fundamental and of too long standing to be attributable solely to any mere bad teaching in the schools.

12. It is exasperating to have to report that although we live in a world so largely determined by the mental quirks and modes of thought of the scientists, there is precious little by way of serious psychological research on their qualities and attitudes. Perhaps the best over-all treatment is by Anne Roe, *The Making of a Scientist* (New York, 1953). For the present purpose we quote Table 8a, p. 148:

Type of imagery chiefly used, and scientific field.

	Visual	*Verbal*	*Total*
Biologists	10	4	14
Exper. Physicists	6	0	6
Theor. Physicists	3	4	7
Social Scientists	2	11	13

The sample is small (only forty case histories in all) and the methods of analysis and definition are not by any means impeccable, but the lopsidedness of the results would indicate that further work on these lines may be worth while. The classical description of the two types of mind is in Henri Poincaré, *The Value of Science* (New York, 1958), Ch. 1. There, citing living mathematicians by name, he makes an appealing case for putting them into watertight compartments as working either by analysis or by geometry and never by both.

Can it be that the Babylonians and Greeks among us do not communicate with one another very well in this sphere where they met only once at a high level? To put it in more psychological terms, we may have here a problem in which we should do well to distinguish between the visualists and the verbal thinkers (if this is the modern equivalent of the old types) and, if we find them distributed bimodally, we should perhaps arrange for each group to have a teacher and a method of the correct mathematical blood group.

If cross-fertilization happens to be vastly more important than may appear at first sight to the scientist, a new evaluation of specialized professional training is called for. Oppressed by the exigencies of a single field that becomes impossibly demanding of his time and energies, the scientist might be wise to specialize much more narrowly than ever, so that he might have enough surplus energy to do something equally near the research front but in a quite different field. Knowledge of two small sectors of the research front might be more effective than knowledge of one sector twice as wide. Since fragmentation is so obviously dangerous, how much of it shall we need? To cover, let us say, about a thousand bits of research front, so that each scientist knew a different pair of bits, would take a million researchers. With less than that, or with duplication of the more popular choices, some borderlands would go unwatched.

Fortunately, the practicability of such drafting of scientists to interdisciplinary fields seems so ludicrously low that we need establish no calculus of field combinations. It might, however, be wise for any embryonic scientist or his advisor to consider the possibility that an unusual diversity in his training (for example, a course in algebraic topology for a biochemist) might be more useful than that which seems more naturally relevant.

To range not quite so far afield, one might also point to the Graeco-Babylonian episode as the supreme example of the value of cross-fertilization in science. If the whole origin of our exact sciences, and hence of our other sciences, is due to a meeting between people who had used methods that were different but applicable to a single interest, how much more important it becomes to make sure that this process may continue. Whole new sciences have arisen as the result of the confluence and interlocking of previously separate departments of knowledge. Historically speaking, many of these have been due more to happy accident than to deliberate planning. Indeed, this is the strongest argument for the unpredictability of research and against the otherwise natural inclination of a society to plan the general direction of its fundamental researches.

I feel that we must attempt to understand the historical processes of such cross-fertilizations a little more clearly and perhaps use such understanding to plan better our scientific education and research facilities so that we may give our scientists all opportunity possible in the teeth of a situation that is tending daily to increase specialization and to decrease the chance that far-flung provinces of science should interact.

POSTSCRIPT

I am surprised that little more has been made of the difference between the styles of thought which have been referred to here as Greek and Babylonian. Though the sharp difference is most apparent in their mathematics and in the historic consequences of these two quite distinct but mutually interacting systems, surely it is also apparent in art and in literature. Think, for example, of the Mayan, Hindu, and Babylonian art works with their clutter of content-laden symbolism designed to be read sequentially and analytically, and compare it with the clean visual and intuitive lines of the Parthenon! Strangely enough, it has now emerged from the psychological researches of Robert

Ornstein and others that the difference in styles corresponds very closely with that of the activity of the left and right hemispheres of the human brain. The left hemisphere, controlling the right half of the body, seems to be "Babylonian," the right hemisphere and left half of the body "Greek." Can it really be that whole civilizations have shown patterns of dominance so pure? If so, a great deal of our modern civilization and, indeed, its special characteristic, must lie in the training of whatever mechanism it is that has led the two separate halves to interact and cross-fertilize each other's creativities either on the individual or the societal level.

CHAPTER 2

Celestial Clockwork in Greece and China

WE WOULD often like to think that our voyages of exploration in the world of learning were precisely navigated or that they followed prevailing winds of scholarship. As often as not, however, it is the chance storm that drives us to unsuspected places and makes us discover America when looking for the Indies.

On some three and a half occasions it has been my extraordinary good luck to have been precipitated into unfamiliar and rich regions where I would never have looked but for the winds of fate that suddenly puffed my sails. The fortunate fact that these several happenings proved coherent provides my excuse for attempting to communicate some of the excitement as well as the conclusions of this personal testament. These researches just "happened," and the only reasonable attitude must be gratitude for circumstances and above all for colleagues whose friendly help enabled a mere trespasser to savor the delights of the medievalist, archaeologist, and Sinologist.

My original course was set, in 1950, toward a study of the experimental tools and laboratories of the scientist, a good borderland area lying between the histories of science and technology. This was in accordance with my specialized training in experimenting with scientific instruments, and it was a particularly appropriate subject at Cambridge University, where the then recently opened Whipple Museum of the History of Science provided access to a wonderful collection of antique instruments exemplifying the only prime documents in that field.

The instruments, and indeed all the available secondary histories, provided reasonably complete documentation only after the sixteenth century, which saw the proliferation of practical science and heralded the Scientific Revolution. From that time on, there was plenty of material to work with. Before that period, sources were remarkably scarce and it was apparent that a considerable effort should be made to see what there was in medieval times and perhaps back into antiquity.

With this in mind, and also being aware of the rare privilege of constant access to the great manuscript collections of Cambridge, I made a point of trying to examine every available medieval book that contained something about scientific instruments. After some months of relatively trivial result, and at a point about halfway through my list of manuscripts to look at, a gust blew for me. At the Perne Library of Peterhouse—the oldest Cambridge library—there was but one noteworthy item dealing with instruments. The catalogue described this as a tract, Latin *incipit* cited, "on the construction of an astrolabe (?)." It was a rather dull volume, traditionally attributed to an obscure astronomer, and it had probably hardly been opened in the last five hundred years it had been in the library.

As I opened it, the shock was considerable. The instrument pictured there was quite unlike an astrolabe—or anything else immediately recognizable. The manuscript itself was beautifully clear and legible, although full of erasures and corrections exactly like an author's draft after polishing (which indeed it almost certainly is) and, above all, nearly every page was dated 1392 and written in Middle English instead of Latin. My high school had had a mad English teacher who, instead of spoiling Shakespeare, taught us Old and Middle English for a year, so fortuitously I was not completely unprepared for the task.

The significance of the date was this: the most important medieval text on an instrument, Chaucer's well-known *Treatise on the Astrolabe,* was written in 1391. To find another English instrument tract dated in the following year was like asking "What happened at Hastings in 1067?" The conclusion was inescapable that this text must have had something to do with Chaucer. It was an exciting chase, which led to the eventually published thesis that this was indeed (very probably) a second astronomical tract by our great poet—and, moreover, the only work in his own handwriting.[1] Perhaps the most hectic part of the sleuthing, I have never dared tell before. It was a search in the Public Record Office to compare the writing on the Peterhouse manuscript with that on a slip of paper which had been proposed as the only other possible document that might be a Chaucer autograph. The slip was one of several dozen, threaded together on a string in a "file" bundle which the Record Office librarian brought. He was on the point of looking in the catalogue to see which of all those was the

1. A preliminary account of this discovery was published in *The Times Literary Supplement* for February 29 and March 7, 1952, and was later reprinted in several versions elsewhere. The final definitive monograph, in which I was assisted by a linguistic analysis by Professor R. M. Wilson of Sheffield, was published as *The Equatorie of the Planetis* (Cambridge, 1955). The tentative ascription to Chaucer has been upheld by most of the scholarly reviewers; this is now also supported by a discovery made by E. S. Kennedy and reported in *Speculum, 34* (1959), 629, that a horoscope in the text was drawn from Messahalla—the source also of Chaucer's *Treatise on the Astrolabe.*

one in question, when I stopped him, riffled through the bundle and immediately saw, standing out dramatically, the one slip that seemed unquestionably in the same hand. It was indeed the very one sought.

By the end of this research I was considerably more familiar with the history and structure of the "planetary equatorium"—the instrument which Chaucer had described as a companion piece to his astrolabe. This pair of instruments was to a medieval astronomer what a slide rule is to an engineer. The astrolabe was used to calculate the positions of the stars in the heavens (it could also be used for simple observations, just as a slide rule can function as a straight edge) and the equatorium was used to calculate the positions of the planets among the stars.

This new background in the early history of other instruments led me to realize that the astrolabe and equatorium occupied a strategic place in history. They were by far the most complicated and sophisticated artifacts throughout the Middle Ages. Their history seemed to extend back continuously in that period, though it was uncertain whether they should be ascribed to a Hellenistic or just an early medieval origin. At the other end of the time scale, they survived in some form or other until the sixteenth and seventeenth centuries, becoming then involved with the great astronomical clocks of the Renaissance and the orreries and planetariums which, respectively, had such a spectacular vogue in the eighteenth and twentieth centuries.

Here one was fishing in very rich waters. The specific task at hand was to see whether the astrolabe and equatorium would contribute to what was surely a very complex and unsatisfactory state of knowledge of the origin of these astronomical showpieces. They heavily influenced the thought of such people as the theologian Paley, the scientist Boyle, and the poets Dante and, of course, Chaucer. They pushed philosophy toward mechanistic determinism. Put in its setting of the history of science, the larger task seemed to be one that was fundamental for our understanding of modern science.

This large task concerns an appreciation of the fact that our civilization has produced not merely a high intellectual grasp of science but also a high scientific technology. By this is meant something distinct from the background noise of the low technology that each civilization and society has evolved as part of its daily life. The various crafts of the primitive industrial chemists, of the metallurgists, of the medical men, of the agriculturists—all these might become highly developed without presaging a scientific or industrial revolution such as we have experienced in the past three or four centuries.

The high scientific technology seems to be based upon the artifacts produced by and for scientists, primarily for their own scientific purposes. The most obvious manifestation of this appeared in the seventeenth century, when all sorts of complex scientific gadgets and instruments were produced and proliferated to the point where they are now familiar as the basic equipment of the modern scientific laboratory; this is, indeed, the story of the rise of modern experimental science. Curiously enough, this movement does not seem to have sprung into being in response to any need or desire on the part of the scientists for devices they might use to make experiments and perform measurements. Galileo and Hooke extended their senses by telescope and microscope, but it took decades before these tools found further application.

On the contrary, it seems clear that in the sixteenth and earlier centuries the world was already full of ingenious artisans who made scientific devices that were more wondrous and beautiful than directly useful. Of course, many of the things, to be salable at all, had to be useful to a point.

Consider, for example, the clock. It certainly had some use in telling the time a little more accurately than common sundials, but one gets much more the impression that even the common domestic clock, not to speak of the great cathedral clock, was regarded in early times more as a marvel and as a piece of conspicuous expenditure than as an instrument that satisfied any urgent practical need. The usefulness, of course, developed later. Eventually the artisans became so clever and were producing such fine products that the public and the scientists came to them to obtain not only clocks but a whole range of other scientific devices.

It seemed, then, that given, let us say, the clockworks of the sixteenth century, one could proceed in reasonably continuous historical understanding to the advanced instruments built by Robert Hooke for the early Royal Society, and from that point by equally easy stages to the cyclotrons and radio telescopes of today's physics laboratories and also to the assembly lines of Detroit. The problem was to account for the production of highly complicated clockwork and the development of its ingenious craftsmen in the sixteenth century.

Now, the history of the mechanical clock is as peculiar as it is fundamental. Almost any book on the history of time measurement opens with a pious first chapter dealing with sundials and water clocks, followed by a chapter in which the first mechanical clock described looks recognizably modern. The beginning is indeed so abrupt that it often seems to me that the phrase "history of time measurement" must have been expressly coined to conceal from the public the awful fact that the clock (as distinct from other time-telling devices) had no early history. It appears to spring forth at birth fully formed and in healthy maturity, needing only a few improvements such as the substitution of a pendulum for the foliot balance and the refinement of the tick-tocking escapement into a precision mechanism.

It is even worse than this. It so happens that the very earliest mechanical clocks we know are the magnificent astronomical showpieces, such as the great clocks of Strasbourg Cathedral and Prague. In fact, the earliest of them all, a clock built by Giovanni de Dondi in Padua in 1364, is by far the most complicated of the series.[2] It contains seven dials, showing each of the planets and all sorts of other astronomical data, with an extra rather inconspicuous dial that tells the time. It uses intricate multiple trains of gear wheels, even with pairs of elliptical gear wheels, link motions, and every conceivable mechanical device. Nothing quite so exquisite mechanically was built again, so far as we know, until a couple of centuries later. Even today a more cunningly contrived piece of clockwork would be hard to find.

If one begins the history of the clock with this specimen, it is plain that the art declines for a long time thereafter, and that a glorious machine that simulates the design of the Creator by making a model of His astronomical universe is eventually simplified into a device that merely tells the time. Thus, one might well regard the modern clock as being nought but a fallen angel from the world of astronomy! What, however, of the state of things before de Dondi? His clock contains the very remarkable device of the escapement and all the wheelwork and weight-drive that is basic to the original invention. Where did these inventions come from? Something so sophisticated as the escapement could not have come into being suddenly except by a stroke of genius,

2. The full texts and illustrations of the Latin manuscripts on the masterpiece by Giovanni de Dondi have never been published in the original or in translation. It is hoped that an edition may be prepared shortly as part of a series of source texts to be published by the University of California. For some years the only reasonable synopsis had been one by H. Alan Lloyd, published without imprint or date (Lausanne, 1955?), but this has now been re-edited in slightly shorter form in the same author's *Some Outstanding Clocks over Seven Hundred Years* (London, 1958), Ch. 3, pt. 1.

and in such a case we might reasonably expect that some hint of the invention should have been preserved. We are, however, completely ignorant of a beginning. All that de Dondi tells us is that the escapement is a common device in his time.

To inject some unity into the story, I therefore attempted to disentangle the clock from the history of time measurement and connect it instead with the longer and earlier history of astronomical models such as the astrolabe and equatorium. Luck was with me, for it seemed just the attitude that was needed. It so happens that all the available examples of geared, clockwork-like, fine mechanical devices before the advent of the clock were models of this sort; we call them "proto-clocks." There were several useful examples, preserved in museums or mentioned in texts, that connected well with this development; they were geared astrolabes and mechanical calculators for the planetary motions, and they seemed to have a quite continuous history.[3]

This led to the tentative hypothesis that the early perfection of astronomical theory had induced men to make divine machines to duplicate the heavenly motions. These proto-clocks were necessarily as complex as the astronomical theory, and their execution called forth a great deal of fine mechanical skill of a sort not expended elsewhere in early times. Such models acted as a medium for the transmission among scientific artisans through the ages of high skills which reached a pinnacle in the late Middle Ages and Renaissance and provided a reservoir of mechanical ability that

3. My findings on this score were published in a pair of articles entitled "Clockwork Before the Clock," which first appeared in *Horological Journal*, *97* (1955), 810, and *98* (1956), 31, and were later reprinted in a polyglot edition (*Germ.* Die Ur-Uhr!) by the *Journal Suisse d'Horlogerie et de Bijouterie* (Lausanne, no date or imprint). A revised and amplified version of this material was embodied in the more accessible monograph, "On the Origin of Clockwork, Perpetual Motion Devices and the Compass," in the series *Contributions from the Museum of History and Technology*, published as *United States National Museum, Bulletin 218* (Washington, 1959).

must be regarded as the source of our later scientific instrumentation.

There were still many problems to solve. Perhaps the
greatest was that of the mysterious origin of the clock escapement, one of the few major inventions that remained
completely anonymous and unaccounted for. While worrying about this, I called one day at the office of Joseph Needham in Cambridge, famous for his monolithic work on
Science and Civilization in China. My purpose was to seek
the latest information on a well-known mechanical equatorium, a planetarium-like object that had been constructed
by Su Sung in A.D. 1088, at the height of the Sung Dynasty
in China. In a sense it is "well-known" because Su Sung's
book, first written in 1092, has been several times reprinted
and republished—most recently in 1922—and has often
been quoted and cited in histories. But those who had written about it, and presumably all those who had looked at
the many editions, had apparently never bothered to read
the really technical material in it or to examine critically
the numerous diagrams showing these mechanical details.[4]

Quite apart from sundry astronomical peculiarities and
the fact that the prime mover looked like a large water

4. This was, however quite true so far as we knew at the time our study
was begun, during 1954, and still true in January 1956, when we first
reported on our findings to the (British) Antiquarian Horological Society.
Only later, in the summer of that year, at the International Congress for
the History of Science at Florence, did we find that colleagues in China
had also been working on Su Sung's clock and had published (in Chinese)
before us. The work had been carried out by Dr. Liu Hsien-Chou, vice-
president of the Ch'ing-Hua University, and in the course of papers on
power sources and transmission in medieval China he had reached the same
conclusions as we had and had published them in October 1953 and July
1954, on both occasions in journals that were not then available in the
West. Our own monographic studies were by this time well advanced and
covered much more ground than that of Dr. Liu, especially in consideration
of the historical significance of the escapement-like device; we therefore
benefited considerably by the discussions with our colleagues and proceeded
with the full publication.

wheel, there was an intriguing arrangement of rods and pivoted bars and levers that seemed from the picture to act as an escapement, checking the motions of the wheels. Now this object was securely dated some three centuries earlier that the first European mention of the escapement, and Needham needed little further urging to translate pieces of the text and confirm that the mechanism was indeed an escapement.

From then on we worked day and night for some four months, with Needham and his assistant, Wang Ling, translating texts and providing the rapidly increasing historical background, so that together we could understand the mechanical details and fit this object into the known history of scientific technology. Thanks to the early invention of printing in China, and to the Chinese custom of producing in each dynasty a sort of analogue to *Great Books of the Western World* so that little of vital importance was lost, we have amazingly fine documentation for Su Sung and his machine. The information preserved is perhaps superior in completeness in some details to the facts we have about many nineteenth- and twentieth-century inventions. The only very considerable difficulty arises from a peculiarity of the Chinese language: the constantly changing and allegorical meanings and nuances of medieval technical terminology, which makes the researcher's task seem like a running crossword puzzle.

Still, thanks to the comprehensiveness of Su Sung's book and the accompanying sources, we were able to work out an exact understanding, almost a modern engineering specification, for his machine. In the course of this we acquired so much new understanding of the terms that we were able to seek other more fragmentary texts and glean from them a previously unintelligible but now usefully complete story of how Su Sung was only the end of a long line of similar people who had built similar devices from the Han dynasty (approximately Roman times) onwards.[5]

Su Sung's great device may be called an astronomical clocktower. It stood some thirty feet high, with another ten feet of observing instruments mounted on a platform on top. Concealed within the housing was a giant water wheel fed by a carefully controlled flow that dripped at a steady rate, filling the buckets of the wheel slowly. Each quarter-hour the wheel became so loaded that it tripped its escapement mechanism, and the whole tower burst into a cacophonous activity with a great creaking and groaning of wheels and levers. On the tower top, the observing instrument was turned automatically to keep pointed steadily at the moving heavens. In a chamber below, a marked star-globe also rotated automatically to provide a microcosm on which the astronomer could see the risings and settings of stars and planets without going outside; it is said proudly that "it agreed with the heavens like two halves of a tally." On the front of the clocktower was a miniature pagoda with a series of doors one above the other. At appointed times, whenever the escapement tripped, these doors would open and little mannikins would appear holding tablets marked with the hours of the day and night, ringing little bells, clashing cymbals, and sounding gongs. It must have been a most spectacular sideshow.

For all the complexity of its externals, the Su Sung clocktower was a comparatively simple mechanism. The big water wheel needed only a simple pair of gears to connect it to the rest of the paraphernalia, which in turn needed only the most elementary mechanical levers and such de-

5. The complete monograph has been published as Monograph No. 1 of the Antiquarian Horological Society, Joseph Needham, Wang Ling, and Derek J. de Solla Price, *Heavenly Clockwork, the Great Astronomical Clocks of Medieval China* (Cambridge, 1960).

vices to produce its effects. Only the escapement mechanism was totally unexpected and refined. It did not tick backwards and forwards quickly, as in the mechanical clock, controlling all the time-keeping properties. Neither was it like the European water clocks, in which a continuous stream of water produced continuous or intermittent action depending solely on the rate of drip of the water. This was definitely an intermediate and missing link in the development. We managed to trace the invention of this form of water-and-lever escapement back to one of the many earlier astronomical clocks built in A.D. 725 by the Tantric monk I-Hsing and his engineering collaborator Liang Ling-Tsan. We also succeeded in tracing the line back to the first known clock in the series, which had been built, perhaps as a non-timekeeping astronomical model, by Chang Hêng, about A.D. 120–140.

What was perhaps more important was that we were able to suggest, at least, how this Chinese invention might have been transmitted to Europe. Curiously enough, one of the other workers on clocks, contemporary with Su Sung, was Shen Kua, who is deeply involved with the history of the magnetic compass. This device seems to have become known in Europe at much the same time as the escapement would have come if it, too, had been transmitted to Europe and was not a home product as we had previously supposed.

Bound up with this is another curiosity. The chimera of perpetual motion machines, well known as one of the most severe mechanical delusions of mankind, seems also to have first become prominent in Europe at this same time; it was quite unknown in antiquity. There are several Latin and Arabic manuscript sources and allusions which involve two or even all three of these otherwise unconnected items, the mechanical clock, the magnetic compass, and the idea of a wheel which would revolve by itself without external power. Time and time again one finds this intrinsically unlikely combination of interests. As yet we have no proof, but I suspect very strongly that all three items emanate from some medieval traveler who made a visit to the circle of Su Sung. Vague tales of the marvelous clock and of the magnetic compass could easily be told in Europe and lead mechanics there to contrive some arrangement of levers that could control the speed of a wheel and make it move round in time with the heavens. Just such a stimulating rumor led Galileo to reinvent the telescope. As to perpetual motion, what is more natural in a traveler's tale after he has seen this giant water wheel inside Su Sung's clock turning without a stream to drive it? How was the traveler to know that each night there came a band of men to turn the pump handles and force the tons of water from the bottom sump to the upper reservoir, thus winding the clock for another day of apparently powerless activity? [6]

In the context of the larger history of civilizations, it is of the greatest interest that heavenly clockwork developed not only in the West but also in China, where mathematical astronomy was much weaker and not nearly so complicated. The reason is, of course, that even something so basic and mathematically simple as the daily cycle of rotation of sun and stars, and the yearly cycle of the sun and calendar, was so fascinating that it must have been almost irresistible for some men to play god and make their own little universe. It bears emphasizing that since the existence of such clockwork is the most sensitive barometer we have for the strength of the high scientific technology in a society, we must say that at this period in the Sung, the Chinese had reached a very remarkable level in the ratio of high technology to pure science. In East and West the technology

6. This unexpected connection between the genesis of clockwork and the idea of perpetual motion machines has now been elaborated in my paper "On the Origin of Clockwork, Perpetual Motion Devices and the Compass" in *United States National Museum, Bulletin 218* (Washington, 1959).

must have been at much the same level, insofar as one can compare them at all. In the East, pure science was certainly not inconsiderable; the Chinese had done many things not yet achieved at that time in Europe. The West, on the other hand, had that special glory of high-powered mathematical astronomy that eventually dominated our scientific destiny.

The more recent events in the chronological development were beginning to fall into a pattern. It provided a whole range of clockwork before the clock, included a reasonable suggestion for the origin of the escapement, and united the previously separate provinces of water clocks, mechanical clocks, and astronomical proto-clocks. One might add that there resulted even more security in the supposition that this was no mere piece of antiquarian parochialism within a province of the history of technology or science. Rather it was an essential key that would lead ultimately from some beginning to an understanding of the whole world of fine mechanics and complicated machines that grew up during the Scientific and Industrial Revolutions. This should be a history with more structure than an almost independent linear series of great inventors and mechanics each with his own special problem.

At this point, taking stock of the situation, I began to feel more puzzled about the historical origins of the whole process at the early end of the time scale. Although I felt sure in my bones that the initial motivation for divine astronomical models must have come from the complex Graeco-Babylonian astronomy in Hellenistic times, there seemed little to support the conjecture. The astrolabe, it is true, was mentioned by Ptolemy and might well have been invented, in principle at least, by Hipparchus in the second century B.C. This is, however, a mechanically very simple device, though mathematically most ingenious. It consists merely of a special circular star-map that may be suitably revolved to show where the stars are at any time of any night of the year. It is still used in modified form (as a "star-finder") by Boy Scouts and others, though the old brass astrolabe with its mathematical elegance of stereographic projection is a more delectable instrument than the cardboard star-finder of today.

What was needed as supporting data was some highly complex mechanical device from antiquity, preferably full of gear wheels and obviously constituting a precursor of the clock. But when one examines Greek mechanical devices critically in a hunt for clockwork, all the ingenuity and appearance of complexity seem to evaporate. Almost our only sources for description of machinery are the writings of Archimedes, Hero of Alexandria, Vitruvius. All these writers mention the use of geared wheels in some form or other, and it seems quite likely that the use of geared wheels must have risen quite early, perhaps around Archimedes' time.

For all the evidence of the use of gear wheels in simple pairs, there appeared not a single example of anything that we would regard as a complex machine. Perhaps the best is the taximeter or hodometer described by both Hero and Vitruvius, but this employed only pairs of gears in tandem to provide a very high ratio for speed reduction. It was a counter that indicated miles traveled by recording the number of revolutions made by a peg on the axle of a carriage or of a special paddle wheel hung over the side of a boat.

If this is the beginning of all clockwork, it is not very glorious, and frankly I hoped for something better, though at my ears was the solemn judgment of the classicists that the Greeks were not interested in these degrading mechanic occupations. There are good authorities for this attitude, and it may be a reasonable consequence of the existence of slavery, as has often been noted. Thus the Greeks appeared to be interested in mechanics only for what mental gymnastics it could afford and preferred to pass silently over as much as possible of the low, everyday technology. There was ground for hope, however, because Hero of Alexandria shows in his book on the *Automaton Theater* and in his *Hydrostatics* a certain schoolboy delight in ingenious trick devices. Though none of these devices uses anything mechanically more advanced than simple levers, strings, and, in a few odd instances, gears, here was the right attitude. This, however, in such weak form, could not be all there was to show for the great days of Greece.

At this point the winds of chance blew me to haven at the Institute for Advanced Study at Princeton, in the company of a number of fine classicists, epigraphers, and archaeologists, as well as physicists and other scientists. In their company it seemed to be natural to bring out of cold storage the one piece of material evidence in this field. It had been considered exciting by all researchers but had hitherto been rejected by all because of difficulties so overpowering that it seemed hopeless to consider it anything but an oddity that we might some day approach when further material came to light. This evidence was an object brought to the surface in the first and unexpected discovery in underwater archaeology in 1900. During that year, Greek sponge divers, driven by storm to anchor near the tiny island of Antikythera, below Kythera in the south of the Peloponnesus, came upon the wreck of a treasure ship. Later research has shown that the ship, loaded with bronze and marble statues and other art objects, must have been wrecked about 65 B.C. (plus or minus ten years) while making a journey from the neighborhood of Rhodes and Cos and on its way presumably to Rome.

Among the surviving art objects and the unrecognizable lumps of corroded bronze and pock-marked marble, there was one pitiably formless lump not noticed particularly when it was first hauled from the sea. Some time later, while drying out, it split into pieces, and the archaeologists on the job immediately recognized it as being of the greatest importance. Within the lump were the remains of bronze plates to which adhered the remnants of many complicated gear wheels and engraved scales. Some of the plates were marked with barely recognizable inscriptions written in Greek characters of the first century B.C., and just enough could be made of the sense to tell that the subject matter was astronomical.

Unfortunately, the effect of two thousand years of underwater decomposition was so great that debris from the corroded exterior hid nearly all of the internal detail of inscription and mechanical construction. In the absence of vital evidence, the available information was published; only rather uncertain and tentative speculation was possible about the nature of the device. In the main, the experts agreed that we had here an important relic of a complex geared astronomical machine, but opinions differed about its analysis and any relation it might have to the astrolabe or to a sort of planetarium that Archimedes is said to have made. Several efforts were made by scholars during the first half of this century, but the matter remained inconclusive and had to stay that way until the painstakingly slow labors of the museum technicians had cleaned away enough debris from the fragments of bronze so that more inscription could be seen and more gear wheels measured.

With my new interest in astronomical machinery, and the facilities and help of the Institute at my disposal, I carefully re-examined a set of new photographs of the fragments which had kindly been provided for me a few years before by the Director of the National Archaeological Museum at Athens. Although a considerable cleaning of the fragments had been effected since the last publication of data, and the lettering and gearwork both seemed much clearer than before, they were not clear enough to make it possible to solve the three-dimensional jigsaw puzzle of fitting fragments together by relying on the photographs alone, and it was obvious that I would need to handle the fragments in order to get any further.[7]

A grant from the American Philosophical Society made it possible for me to visit Athens that summer, and through capricious and fortunate circumstances, the assistance was available there of George Stamires, an epigrapher friend from the Institute, who helped me by masterly readings of the difficult inscriptions. The museum authorities were most cooperative, and it proved a none too arduous task to sketch all the interconnections and details of the wheels within the mechanism, measure everything that could be measured, and photograph every aspect of every little fragment. So armed, I returned eventually to Princeton and to the jigsaw puzzle.

Little by little the pieces fitted together until there resulted a fair idea of the nature and purpose of the machine and of the main character of the inscriptions with which it was covered. The original Antikythera mechanism must have borne remarkable resemblance to a good modern me-

7. I have now published a popular and tentative account of the Antikythera fragments in *Scientific American,* 200 (June, 1959), 60–7, and a short and formal statement with bibliography in *Year Book of the American Philosophical Society* (1959), pp. 618–19. Since these publications, the matter of date and provenance of the wreck at Antikythera has been reported upon by G. R. Edwards at the American Institute of Archaeology, December 30, 1959. I am most grateful to him for a typescript of this yet unpublished address. The conclusion is that the ship set forth on a commercial voyage, carrying sculpture consigned from an Aegean source, probably for the Italian market, in the early second quarter of the first century B.C. I have, as well, had access to unpublished notes and photographs made by the late Professor Albert Rehm, who worked for many years on this material without a full publication of his findings; the information, kindly made available to me by the Bayerische Staatsbibliothek in Munich, has confirmed and at some points extended the previous findings. See also pp. 47–48.

chanical clock. It consisted of a wooden frame which supported metal plates front and back, each plate having quite
complicated dials with pointers moving around them. The
whole device was about as large as a thick folio encyclopedia
volume. Inside the box formed by frame and plates was a
mechanism of gear wheels, some twenty of them at least,
arranged in a non-obvious way and including differential
gears and a crown wheel, the whole lot being mounted on
an internal bronze plate. A shaft ran into the box from the
side, and when this was turned all the pointers moved over
their dials at various speeds. The dial plates were protected
by bronze doors hinged to them, and dials and doors carried the long inscriptions which described how the machine
was to be operated.

It appears that this was indeed designed as a computing
machine that could work out and exhibit the motions of
the sun and moon and probably also the planets. Exactly
how it did it is not clear, but the evidence thus far suggests
that it was quite different from all other planetary models.
It was not like the more familiar planetarium or orrery,
which shows the planets moving around at their various
speeds, but much more like a mechanization of the purely
arithmetical Babylonian methods. One just read the dials
in accordance with the instructions, and legends on the dials
indicated which astronomical phenomena would be happening at any particular time.

The antiquarian detail of this investigation proved particularly exciting. It was possible from the calendar inscribed on one of the dials to deduce the possibility that the
mechanism had been constructed in 87 B.C. and used for
just two years, during which time it had had two repairs.
Thus it seems likely that it was not more than thirty years
old, certainly no antique, when put aboard the ship. Almost certain too is the evidence that this was no navigating

device used on board ship, as once had been thought; it was, rather, a valuable art object taken, like the rest of the treasure, as booty or as merchandise.

More important was the observation that certain technical details of the construction—the shape of the gear teeth and the general character of the design of gear trains—showed significantly close affinity with the series of medieval Islamic and European proto-clocks, the specimens of astrolabes and equatoria, that had already been attested. Clearly the Greek machine must be neither a freak nor an isolated specimen. It is the first specimen in the line and the hoary primeval ancestor of all clocks, calculating machines, and other abstruse fine mechanical devices.

The existence of this most complex Antikythera mechanism necessarily changes all our ideas about the nature of Greek high technology. We no longer need believe the expressions of a distaste for manual labor but may regard them merely as a very human personal preference of those philosophers whose tastes were otherwise inclined. Hero and Vitruvius should be looked upon as chance survivors that may not by any means be as representative as hitherto assumed.

The problem, in essence, seems rather like that of the little green men who might come from space in A.D. 4000 and find the earth a charred waste, with only a corner of the deep vault of the National Gallery of Art remaining as a sign of Man's reign. Perhaps, considering the Parthenon, we might grant them ruins of a few buildings in academic gothic and a sprinkling of Frank Lloyd Wright. Can you see their reconstruction of humanity, based on these together with a couple of Van Gogh's, a Rembrandt, a Rubens, and three Picassos? But this, notwithstanding, is what we habitually do for ancient civilization. Is it not possible that just as today's artists do not customarily paint electrons and nuclear physics or the design of automobile engines, the

Greek writers did not have the tradition of writing about their machines and sciences unless such writings could constitute a monument of thought?

Whatever the reason for the unexpected *volte-face* after centuries of classical scholarship, we must live with it. In its narrowest implications at least, within the field of clockwork and origins of high technology, the picture now makes much better sense and presents less anomaly. No longer do we need seek some historical reason for the fact that the Chinese built their great clocktowers while the Greeks with more scientific advance in astronomical theory did so little mechanically. The reasonable and expected balance has been found, but the price paid for it must be an antedating of the problem. An object so incredibly complex as the Antikythera mechanism cannot possibly have been the first of its line. More probably its existence lends substance to the bare information we have from Cicero and others about a planetary model made by Archimedes. If true, this is assuredly near the source of the Hellenistic trail that stands at the entrance to the world of scientific machinery.

With three lucky occasions reported, there remains one half-gust of fortune to tell. In 1958, while I was working on the medieval texts dealing with Islamic clocks and machines in this series, a footnote in a modern book revealed to me that one of these clocks still survived some years ago in its original setting, in a high room in the minaret of Karaouyin University mosque in the city of Fez in Morocco. Being at that time in Washington, I happened to meet one of the ministers from the Moroccan Embassy and mentioned the matter to him. In due time, there arrived a set of photographs of this room and full permission to be one of the first of the unbelievers to be permitted to have access to it and to study its contents.

To my surprise, the photographs showed that the old clock appeared to be quite intact in its original state from the fourteenth century. It even indicated the same hour as the other timepieces in the collection! It is, in fact, the oldest working clock in the world. Furthermore, its design, never completely described in print or by any travelers, appeared to be in keeping with everything belonging to a conservative tradition of astronomical water clocks harking back to Hellenistic times, possibly prior to the Antikythera mechanism. The room of the mosque is otherwise simply littered with astrolabes, clocks, and other time-keeping devices ancient and modern. Apparently each official time-keeper of the mosque for centuries back added the best instruments of his day to the collection, and it stands now as a veritable storehouse of antiquarian treasures, amply sufficient to provide enough research material for several lifetimes.

My trip to Morocco must belong to a future summer, when perhaps more pursuit and lucky discoveries will add to the present embarrassment of riches. Certainly the subject is in so primitive a state that each question solved raises more problems than can be handled by the absurdly few workers in the field.

Perhaps it is presumptuous to look the gift horse in the mouth and add the hope that these fortuitous researches might also have established accidentally a crucial phase in the general history of science and technology. Yet this line, which starts with Archimedes and finishes in any modern laboratory, seems vital to the origin of experimental method. That may be considered one leg of science, the other being its Graeco-Babylonian heritage of logic and mathematics. We might not yet know what made science run in the era of Newton, but on these two legs it surely began a sturdy walk.

POSTSCRIPT

The Moroccan researches have now been reported in my

paper, "Mechanical Water Clocks of the 14th Century in Fez, Morocco," *Actes du Xe Congres International d'Histoire des Sciences,* (Ithaca, 1962; Paris: Hermann & Cie, 1964), 1:532–35. The technology of the clocks, though Islamic in workmanship, was purely Hellenistic in conception and preserved a great deal of the detail that we know otherwise only from somewhat vague texts. Their study gave me a very strong feeling that there was a continuity of tradition from the earliest times through Islamic and Christian medieval clock-building.

There is much more progress to report on this clock story. With Joseph Noble I was able to study and complete a reconstruction of the elaborate showpiece mechanism that once graced the interior of the Tower of Winds (see p. 78). Lastly, thanks to a very lucky break with a newly available technique and with the cooperation of Dr. Ch. Karakalos of the Greek Atomic Energy Commission, we were able to obtain radiographs that showed all the mechanism hidden within the corroded fragments of the Antikythera machine. Twenty years after starting work on this most enigmatic of all scientific artifacts of antiquity, I have been able to publish a complete elucidation of the mechanism and the workings of its complicated and highly sophisticated gear trains ("Gears From the Greeks, The Antikythera Mechanism—A Calendar Computer from ca. 80 B.C." in *Transactions of the American Philosophical Society,* vol. 64, pt. 7 (Philadelphia, December, 1974). It turned out that the preliminary guesses were substantially correct and that this was a mechanized version of the Metonic calendar cycle, giving the places of the sun and moon as well as the risings and settings of the circuit of notable fixed stars through the cycles of years and months.

Moreover, in putting together the historical record, it appeared that just such a machine had been seen at exactly this time by the Roman orator Cicero, who was then in

Rhodes for a couple of years studying with various teachers, including the astronomer Posidonios who, he says, had caused such a planetarium machine to be built in the tradition of Archimedes. There is even a possibility that the Antikythera treasure represents the baggage of Cicero, being sent home to Rome after his stay in Rhodes; fortunately, it seems unlikely that we will ever have evidence to support or reject such a conjecture; it would be too good to be true. It is perhaps more important that this device turns out to be very much in the historical tradition we should have expected from the stories about Archimedes; but the sophistication of those gear trains, including the differential gear system, requires us to completely rethink our attitudes toward ancient Greek technology. Men who could build this could have built almost any mechanical device they wanted to. The Greeks cannot now be regarded as great brains who disdained manual labor or rejected technology because of their slave society. The technology was there, and it has just not survived like the great marble buildings, statuary, and the constantly recopied literary works of high culture.

CHAPTER 3

Automata and the Origins of Mechanism and
Mechanistic Philosophy

HISTORIANS of the mechanistic philosophy customarily proceed from the reasonable assumption that certain theories
in astronomy and biology derived from man's familiarity
with various machines and mechanical devices. Using everyday technological artifacts, one could attempt with some
measure of success to explain the motions of the planets
and the behavior of living animals as having much of the
certainty and regularity reproduced in these physical models. Indeed, the steady advancement of technology and the
increase in familiarity with machines and their fundamental theory is usually cited as the decisive factor in the
growth of mechanistic philosophy, especially toward the
beginning of the instrument-dominated Scientific Revolution in the sixteenth and seventeenth centuries.

It seems clear that any interpretation of the interaction
between the histories of technology and philosophy must
assign a special and nodal role to those peculiar mechanisms
designed by ingenious artificers to simulate the natural uni-

verse. In this light we shall now examine the history of such simulacra (i.e. devices that simulate) and automata (i.e. devices that move by themselves), whose very existence offered tangible proof, more impressive than any theory, that the natural universe of physics and biology was susceptible to mechanistic explication.

It is the burden of this chapter to suggest further that in the history of automata is found plain indication that the customary interpretation puts the cart before the horse. Contrary to the popular belief that science proceeds from the simple to the complex, it seems as if mechanistic philosophy—or *mechanicism,* to use the appropriate term coined by Dijksterhuis[1]—led to mechanism rather than the other way about. We suggest that some strong innate urge toward mechanistic explanation led to the making of automata, and that from automata has evolved much of our technology, particularly the part embracing fine mechanism and scientific instrumentation. When the old interpretation has been thus reversed, the history of automata assumes an importance even greater than before. In these special mechanisms are seen the progenitors of the Industrial Revolution. In the augmenting success of automata through the age of Descartes, and perhaps up to and including the age of electronic computers, we see the prime tangible manifestation of the triumph of rational, mechanistic explanation over those of the vitalists and theologians.

Our story, then, begins with the deep-rooted urge of man to simulate the world about him through the graphic and plastic arts. The almost magical, naturalistic rock paintings of prehistoric caves, the ancient grotesque figurines and other "idols" found in burials, testify to the ancient origin of this urge in primitive religion. It is clear that long before the flowering of Greek civilization man had taken his first

1. E. J. Dijksterhuis, *The Mechanization of the World Picture* (Oxford, 1961), 3*n.*

faltering steps toward elaborating pictorial representations with automation. Chapuis[2] points to the development of dolls with jointed arms and other articulated figurines such as those from ancient Egyptian tombs (from the Twelfth Dynasty onward) and takes these as proto-automata. Interestingly enough, it is from just such figurines as these, representing scenes of battle, in ships, in bakeries, and so on, that the modern historian of technology often obtains his most valuable information about the crafts and everyday life of deep antiquity.[3]

Perhaps the next level of sophistication is also found in ancient Egypt: talking statues worked by means of a speaking trumpet concealed in hollows leading down from the mouth. Two such statues are extant: a painted wooden head of the jackel God of the Dead is preserved in the Louvre, and a large white limestone bust of the god Re-Harmakhis of Lower Egypt is in the Cairo Museum and was described in technical detail by Loukianoff.[4] Jointed and talking figures are not confined to Egypt but probably occurred early in civilization and are widespread. The articulated masks to be worn over the face, found in Africa, and the famous *Wayang* figures of flat, jointed leather for traditional Indonesian shadow plays are pointers in this direction. Primitive animism may lie at the very root of animation.

It seems that by the beginning of Greek culture the process of natural exaggeration in mythology and legend had produced at least the concept of simulacra able to do more than merely talk and move their arms. Daedalus, as well as imitating the flight of birds, is said (ps. Aristotle,

2. Alfred Chapuis and Edmond Droz, *Automata* (New York, 1958), pp. 13–29.

3. See Charles Singer et al., eds., *A History of Technology*, vol. 1 (Oxford, 1954), pp. 427, 437, plate 13 A.

4. Gregoire Loukianoff, "Une statue parlante, ou Oracle du dieu Re-Harmakhis," *Annales du service des Antiquités de l'Egypte* (Cairo, 1936), pp. 187–93.

De Anima, i, 3) to have contrived statues that moved and walked in front of the Labyrinth, guarding it; and Archytas of Tarentum (fourth century B.C.) is said to have made a flying dove of wood worked by counterweights and air pressure. That such a tradition, supported by devices probably no more complex than those of ancient Egypt, was taken seriously, is indicated by the use of moving statues to deliver oracles and by their later Roman counterpart, the *neuropastes.*

An impressive use of an automaton resulted from the murder of Julius Caesar. On the day of Caesar's funeral, the city of Rome was the scene of great confusion and tumult over the death of its idol. Mark Antony was to deliver the funeral oration, and he was determined to arouse the populace to take action against the conspirators. The scene is vividly described by Walter:

> An unendurable anguish weighed upon the quivering crowd. Their nerves were strained to the breaking point. They seemed ready for anything. And now a vision of horror struck them in all its brutality. From the bier Caesar arose and began to turn around slowly, exposing to their terrified gaze his dreadfully livid face and his twenty-three wounds still bleeding. It was a wax model which Antony had ordered in the greatest secrecy and which automatically moved by means of a special mechanism hidden behind the bed.[5]

This realistic automaton did as much as Mark Antony's words to create a riot of the populace at the funeral, which contributed to one of the greatest revolutions in history.

As a literary and imaginative theme, the simulacrum or statue that comes magically to life without mechanical intervention has been with us from the early legends of

5. Gérard Walter, *Caesar: A Biography* (New York, 1952), p. 544.

Vulcan and Pygmalion to the medieval Golem of Jewish folklore, the Faustus legend, the affair of Don Juan's father-in-law, and several miraculous animations of holy images. A variant tradition, in which animation is secured by scientific but non-mechanical artifice, is seen in the homunculus of Paracelsus, which was to be hatched alchemically from a basis of semen nourished by blood, and in the monster of Frankenstein, in which lightning supplied the electric vital fluid.

Although there seems to have been a continuous and strong tradition leading man to simulate living animals and even man himself, in early Greek times the technological skill to materialize this dream more extensively than in speaking tubes and simple jointed arms did not exist. Perhaps the most crucial point in the early history of automata is that this skill seems to have been acquired in the search for a different variety of automaton, the astronomical model.

The prehistory of cosmological simulacra is apparently less extensive than that of biological models, and is later in appearing. Coming to grips with the basic astronomical phenomena probably required a level of sophistication considerably higher than was necessary for a reasonably basic appreciation of the movement of living things. At all events, the beginnings of astronomical representation may be seen in the famous star-map ceilings of Egyptian tombs, the flower-pot shaped clepsydras with their celestial ornamentation, and in the goddess Nut arching over the celestial vault and providing primitive mechanism for the disappearance of the Sun by swallowing it at the end of the day's journey. Perhaps it is not altogether fanciful to see the astronomical zodiac as the first primitive coming together of a cosmic model and a set of animal models.

In the Babylonian area representations of the celestial bodies and the beginnings of a primitive pictorial notion of the structure of the universe are also found. From that same area, moreover, comes the highly sophisticated but non-pictorial mathematical astronomy that achieved the first spectacular success of scientific prediction, a prediction based upon an acute sensitivity for the pattern of natural number rather than for any perception of mechanism or even of geometrical form, but nonetheless deterministic in its findings of precise and regular order in the most common astronomical phenomena.

Babylonian theory was exquisitely complicated and was probably never understood in its entirety by any Greek, but the basic principle of mathematical regularity and the fundamental parameters of the motions could easily be comprehended and transmitted, so that they formed a secure foundation for much of the science which flourished in that age still called "the Greek Miracle." The almost mystic dominance of a regularity of number, which led to Pythagoreanism, and the rationality of celestial motions, translated from Babylonian form into the geometrical imagery of the Greeks, formed the basis of all Hipparchan and Ptolemaic astronomy and much of Greek mathematics. By the time of Plato it seems likely artifacts existed, perhaps even with simple animation, simulating the geometrically understood cosmos. Brumbaugh[6] has pointed out that much of Plato's imagery seems to derive from models that were more than mental figments. Certainly by the time of Eudoxos (ca. 370 B.C.) we find a geometrical model of planetary motion having every appearance of relation to an actual mechanism of bronze rings.

Perhaps the most telling evidence is found in the writings of Ctesibius (300–270 B.C.), who lived, as did Straton the physicist, in the period between Aristotle and Archimedes.

6. Robert S. Brumbaugh, "Plato and the History of Science," *Studium Generale*, vol. 14 (1961), pp. 520–27.

Thanks to the monumental labors of A. G. Drachmann,[7] we now know that the basic mechanisms of water-clocks and other devices familiar from the writings of Vitruvius (ca. 25 B.C.) and Heron of Alexandria (ca. 62 A.D.) go back to this time. Recent excavations at the Agora of Athens and at Oropos confirm the existence there of monumental water-clock edifices from at least the third century B.C. onward; that little gem of architecture, the Tower of Winds in the Roman Agora of Athens, built by Andronicus Cyrrhestes ca. 75 B.C., agrees so well with this theory and is so perfectly preserved (except for the centerpiece mechanism) that from it one can essay a reconstruction of this entire class.

It would be a mistake to suppose that water-clocks, or the sundials to which they are closely related, had the primary utilitarian purpose of telling the time. Doubtless they were on occasion made to serve this practical end, but on the whole their design and intention seems to have been the aesthetic or religious satisfaction derived from making a device to simulate the heavens. Greek and Roman sundials, for example, seldom have their hour-lines numbered, but almost invariably the equator and tropical lines are modeled on their surfaces and suitably inscribed. The design is a mathematical tour de force in elegantly mapping the heavenly vault on a sphere, a cone, a cylinder, or on specially placed planes. The water-clocks, powered by the fall of a float in a container filled or emptied by dripping water (as in the Egyptian clepsydras), not only indicated the time by means of scale and pointer. At first they seem to have been used to turn a simple model of the sun around with a celestial sphere; certainly this was the earliest type of model known in the analogous development in the

7. A. G. Drachmann, "Ktesibios, Philon and Heron; A Study in Ancient Pneumatics," *Acta Historica Scientiarum Naturalium et Medicinalium,* vol. 4 (Copenhagen, 1948), and *The Mechanical Technology of Greek and Roman Antiquity* (Copenhagen; Madison, Wis.; and London, 1963).

Chinese cultural area.[8] Later, presumably by the time of
Hipparchus, the principle of stereographic projection provided a flat map of the heavens bearing the sun and moon
which could be turned to display most impressively an arti-
ficially rotating sky; this device was later adapted into the
astrolabe, the most important of all medieval scientific instruments of computation.[9]

From the evidence of the Tower of Winds, these monumental public structures contained much more than the
astronomical model and its powering clepsydra. Ingenious
sundials were added all around the octagonal tower, and
on top was a bronze Triton weather-vane which pointed to
eight relief figures personifying the winds, mounted on a
frieze surrounding the top of the building. Within the
structure, around the walls, were probably mounted para-
pegma calendars on which were tabulated daily astronomical and meteorological events, events that could be
confirmed visually from the central astronomical showpiece
and from the weather-vane.

Judging from the texts of Heron, Philon, and Ctesibius
collected by Drachmann; from the tradition of automatic
globes and planetaria made by Archimedes; and from the
few extant objects (on which I have previously commented
elsewhere) ;[10] we may say that the technology of astronomical automata underwent a period of intense development.
The first major advances seem to have been made by Ctesibius and Archimedes, and the subsequent improvement

8. Joseph Needham, Wang Ling, and Derek J. de Solla Price, *Heavenly
Clockwork* (Cambridge, 1959).

9. Derek J. de Solla Price, "Precision Instruments to 1500," Ch. 22;
"The Manufacture of Scientific Instruments from c 1500 to c 1700," Ch.
23, in Singer et al., *A History of Technology*, vol. 3 (Oxford, 1954).

10. Derek J. de Solla Price, "On the Origin of Clockwork, Perpetual
Motion Devices and the Compass," *United States National Museum Bul-
letin 218: Contributions from the Museum of History and Technology*,
Paper 6 (Washington, D.C., 1959), pp. 81–112.

must have been prodigious indeed, seeing that it made possible, by the first century B.C., the Antikythera mechanism with its extraordinary complex astronomical gearing.[11] From this we must suppose that the writings of Heron and Vitruvius preserve for us only a small and incidental portion of the corpus of mechanical skill that existed in Hellenistic and Roman times.

Even though we know so little about this sophisticated technology, only, indeed, that preferred part of it that was committed to writing and copied into preservation, its characteristics are obvious—so obvious, that I am surprised previous scholars have not drawn the inevitable conclusions. Amongst historians of technology there seems always to have been private, somewhat peevish discontent because the most ingenious mechanical devices of antiquity were not useful machines but trivial toys. Only slowly do the machines of everyday life take up the scientific advances and basic principles used long before in the despicable playthings and overly ingenious, impracticable scientific models and instruments.

We now suggest that from Ctesibius and Archimedes onward we can see the development of a fine mechanical technology, originating in the improvement of astronomical simulacra from the simple spinning globe to the geared planetarium and anaphoric clock. Partly associated with and partly stemming from these advances, we see the application of similar mechanical principles to biological simulacra. We suggest that these two great varieties of automata go hand-in-hand and are indissolubly wedded in all their subsequent developments. In many ways they appear mechanically and historically dependent upon one other; they represent complementary facets of man's urge to exhibit the depth of his understanding and his sophisticated skills by

11. Derek J. de Solla Price, "An Ancient Greek Computer," *Scientific American* (June 1959), pp. 60–67.

playing the role of a do-it-yourself creator of the universe, embodying its two most noble aspects, the cosmic and the animate.

In support of the thesis that astronomical clockwork and biological automata are complementary to each other, the following evidence is submitted: (a) both types of simulacra see their first extensive development at the same time; (b) the techniques used are found at first only in them, seeping slowly, and much later on, into other instruments and machines; and (c) throughout the entire medieval, Renaissance, and even modern evolution of fine mechanism, a central role is played by great astronomical clocks whose principal characteristic is the combination of astronomical showpiece with the automatic jackwork of imitation animals and human beings.

In Graeco-Roman times the deepest complementarity exists between the clepsydra principles used in astronomical models and clocks, and the almost identical inner workings of the Heronic singing-birds and other *parerga*. Less closely related but still significant are the statuettes holding indicating pointers on the scales of the water-clocks and the Triton figurine with wind gods surmounting the tower of Cyrrhestes. It may be significant that Rhodes, which was a center of astronomy in the first century B.C., and Delos, which manufactured sundials like a Greek Switzerland, were both famed for their automatic statues; even the Colossus of Rhodes is said by Pindar to have been animated in some way.

But since by now we strongly suspect that we know only a fragment of the original fine mechanical tradition of classical times, let us turn next to the Middle Ages. One may reasonably suppose that later examples often preserve, with little refinement, an ancient source. The ample evidence of many well-edited texts and a couple of extant instruments testifies to the existence of a more or less continuous, and remarkably homogeneous, tradition of mechanical waterclocks, mainly from Islam but extending without change to contemporary Byzantium,[12] and with some modifications even as far as China and perhaps India. This tradition seems to have been transmitted to Europe without much change or dilution during the medieval renaissance of the twelfth and thirteenth centuries; during the next century it became conflated with other lines of development and was thus transformed into the essentially modern principle of the mechanical clock, which preserves so much of the feeling and motivation of the old ideal. In particular, there was preserved the special complementary relation between the clockwork and jackwork.

In the typical Islamic clock, which was in its heyday from about 800 A.D. to 1350 A.D. and which may be very close to the lost Hellenistic originals, power is provided by a float in a vessel filled or emptied by dripping water. This power is harnessed, either directly by having a chain or string pull a block along a straight channel, or rotationally by having the string wind around a pulley, or by using a geared pinion and rack. The straight motion may trip a series of levers one by one, opening a set of doors, moving a set of figurines, or letting a series of balls fall into gongs and sound at set intervals. The circular motion may be used to animate automata, moving their heads or bodies or rotating their eyeballs, or to turn a globe or stereographic map of the heavens and perhaps also, by appropriate gearing, models of the sun and moon placed upon the heavenly representation. In a refinement, the dripping water may be caught in another vessel which is suddenly and periodically emptied by an automatic syphon or a balancing-jar; the apparatus then works rather like a faulty modern lavatory cistern

12. Note also the traditional Heronic jackwork described by Gerard Brett, "The Automata in the Byzantine 'Throne of Solomon,'" *Speculum*, vol. 29 (July 1954), pp. 477–87.

that flushes itself as soon as the tank is full. The ensuing
rush of water may then spin a water wheel to move other
automata or it may enter a vessel, displacing air so as to
blow the whistle or sound the organ pipes that provide the
singing of the mechanical birds or other manikins.

These mechanisms, though undoubtedly impressive, are
mechanically simple and Heronic. They are described in
detail by Ridwān and al-Jāzari[13] (both early thirteenth
century), and there are texts describing their appearance in
Damascus and Gaza.[14] There is also evidence of two fairly
simple clocks (fourteenth-century) of this type extant in
Fez, Morocco,[15] and of a quite complex geared astrolabe
designed by al-Biruni ca. 1000 A.D.[16] and attested by an example made in Isfahan in 1221 A.D.[17] We know also that
such devices were reputedly owned by Harun al-Rashid and
Charlemagne[18] in the ninth century A.D., and by Saladin in
the twelfth century.

Just before the transmission to Europe in the thirteenth
century of the corpus of knowledge about clockwork and
automata, that learning somehow became intertwined with
concepts of perpetual motion (an idea apparently unknown
in Classical antiquity) and of magnetism and the mystery

13. Eilhard Wiedemann and Fritz Hauser, "Über die Uhren im Bereich
der islamischen Kultur," *Nova Acta Abh. der Kaiserl. Leop.-Carol.
Deutschen Akademie der Naturforscher,* vol. 100, no. 5 (Halle, 1915).

14. H. Diels, "Über die von Prokop beschriebene Kunstuhr von Gaza,"
Abh. der Königlich Preussischen Akademie der Wissenschaften, Philos-
Hist. Klasse, no. 7 (Berlin, 1917).

15. Derek J. de Solla Price, "Mechanical Water Clocks of the 14th
Century in Fez, Morocco," published in the *Proceedings of the Xth
International Congress of the History of Science* (Ithaca, N.Y. and Phila-
delphia, 1962).

16. E. Wiedemann, "Ein Instrument das die Bewegung von Sonne und
Mond darstellt, nach al Biruni," *Der Islam,* vol. 4 (1913), pp. 5–13.

17. Price, "On the Origin of Clockwork . . . .," *loc. cit.,* pp. 98–100.

18. Note also the interesting astronomical model described in F. N.
Estey, "Charlemagne's Silver Celestial Table," *Speculum,* vol. 18 (1943),
pp. 112–17.

of magnetic force. This intertwining may have originated with the accounts of travelers returned from China telling about the clocks of Su Sung and the related work on the magnet being done there.[19] Also toward the end of the thirteenth century came the purely astronomical elaboration of complicated equatoria; these were designed to compute the positions of planets and afforded a more complete geometrical simulation of Ptolemaic theory than the older, somewhat Aristotelian models embodying simple uniform rotation. Many such devices are seen in the Alfonsine corpus, which also contains designs for a rotating drum with leaky compartments filled with mercury that acts as the regulatory agency of an astrolabe clock. Elsewhere in the Islamic sources are the elements of the weight drive, used not for the mechanical clocks to which it was later adapted, but for pumping water.[20]

With the transmission to medieval Europe of all these ideas and techniques seems to have come a burst of exuberant interest that was further stimulated by the flowering of the craft guilds. The drawings in the famous notebook of Villard de Honnecourt (ca. 1254)—more likely the album of a guild rather than the work of an individual—show a clocktower for a mechanism that was probably a clepsydra-driven bell chime. Other drawings show a simple rope-and-pulley apparatus for turning an automaton angel (which is interpreted quite unauthoritatively as an escapement by Frémont[21]) and another rope-driven automaton bird. Also from the thirteenth century are ample records and even illuminations showing church water-clocks; there is the preoccupation with perpetual motion of Robertus Anglicus in his search for an astronomical simulator, and a similar

19. Price, "On the Origin of Clockwork," pp. 108–10.

20. Hans Schmeller, "Beiträge zur Geschichte der Technik in der Antike und bei den Arabern," *Abh. zur Geschichte der Naturwissenschaften und der Medizin,* no. 6 (Erlangen, 1922).

21. C. Frémont, *Origine de l'horloge à poids* (Paris, 1915).

preoccupation "solved" through magnetic power by Peter Peregrinus.[22]

By 1320 the clock, presumably a water-clock, has been adapted by Richard of Wallingford to the working of complicated automata based on the principle of the equatorium and demonstrating with great ingenuity the exact motions of Ptolemaic astronomy. Not long afterward, in 1364, Giovanni de Dondi had built his great clock in Padua; we know from a full manuscript description that this was a true mechanical clock with weight-drive, verge-and-foliot escapement, seven magnificent dials with a panoply of elliptical and normal gear-wheels and linkwork to show all the astronomical motions, a fully automated calendar showing Easter and other holydays, and—a little dial for telling the time. The clock of de Dondi, though matching in complexity and ingenuity any seventeenth-century product of the clockmaker's art, is somewhat anomalous in our history, for it has no biological jackwork. However, we know that this was a firm tradition by then, for it appears in the first monumental astronomical clock of the cathedral of Strassbourg. From this most famous and influential series of three successive clocks (1354, 1574, 1842) has been most fortunately preserved (in the local museum) the large bronze automated cock which surmounted the structure. Crowing and moving most naturalistically on the hours, the cock accompanied with its actions the carillon, the other manikins, and the astrolabe dial and calendar work. By this time mechanical ingenuity was able to produce automation of the bird figure; the complicated arrangement of strings and levers became a reasonable simulacrum for the musculature and skeleton of a real bird.[23]

22. See Lynn White, *Medieval Technology and Social Change* (Oxford 1962), pp. 120–29, 173.

23. Alfred Ungerer, *Les Horloges Astronomiques et Monumentales les plus remarquables de l'antiquité jusqu'à nos jours* (Strassbourg, 1931), pp. 163–65.

From this time forward, the great astronomical cathedral clocks, complete with jackwork, swept Europe, growing in number but perhaps lessening in mechanical complexity during the sixteenth, seventeenth, and eighteenth centuries. The only interruption occurred during that remarkably dead period of intellectual and economic depression in the second quarter of the fifteenth century. Apart from this, one can trace the steady evolvement of the clockmakers' fine metal-working craft to its finest manifestation in the craft of the instrument-maker which was to dominate the development of learning during the Scientific Revolution.

Accompanying the European popularization of water-clocks and mechanical clocks during the Middle Ages came a flood of literary allusion based partly upon the clocks, partly upon travelers' tales of parallel traditions of technology in Constantinople and the Orient, and partly upon a revival of the classical mystique of magically animated figurines.[24] The clock itself, in its debasement from astronomical masterpiece to mere time-teller, becomes so familiar that is assumes allegorical significance in such disquisitions as Froissart's *L'Horloge Amoureuse* and in the tract *L'Horloge de Sapience,* whose manuscript illuminations have offered recent scholarship so much detailed insight into early mechanics and instrument-making. From Heronic sources, perhaps Byzantine, perhaps transmitted through Arabic to medieval Europe, come many allusions to brass trees full of singing birds, set in motion by water power, by the wind, or by bellows.

More magically still, Albertus Magnus (like many other philosophers) is said to have made a brazen head, and he especially is credited with the feat of having constructed a mechanical man—a *robot,* to use the term coined by Capek —from metal, wax, glass, and leather. We know no specific

24. Merriam Sherwood, "Magic and Mechanics in Medieval Fiction," *Studies in Philology,* vol. 44 (October 1947), pp. 567–92.

details of any such automaton made by Albertus, but we may suppose that at about this period the art of automaton-making in Europe had recovered a level of sophistication and verisimilitude probably not much inferior to that demonstrated in the Strassbourg clock.

Albertus's most famous pupil, St. Thomas Aquinas, stated emphatically in his *Summa Theologica* (Qu. 13, Art. 2, Reply obj. 3, Pt. II) that animals show regular and orderly behavior and must therefore be regarded as machines, distinct from man who has been endowed with a rational soul and therefore acts by reason. Surely, such a near-Cartesian concept could only become possible and convincing when the art of automaton-making had reached the point where it was felt that all orderly movement could be reproduced, in principle at least, by a sufficiently complex machine. It is remarkable that at this very time figures of apes become popular as automata—they had been used *inter alia* by the Islamic clockmakers—being endowed with an appearance similar to that of man but having as a "beast-machine." This is probably the line that led to such literary and philosophical devices as the Yahoos of Jonathan Swift, beasts shaped like men but without rationality; it is also the line that made philosophically important the emergent possibility of exhibiting mechanically many manifestations of apparent rationality.

Of such kind were the mathematical calculating machines that began with all the early astronomical automata, proliferated during the sixteenth and seventeenth centuries, and culminated in the first true digital computer of Pascal, the *Pascaline* of 1645. Of such kind were the remarkably constructed musical automata during the same period, particularly such impressive devices as that built by Achilles Lagenbucher of Augsburg in 1610; this seems to have had a large array of instruments that were programmed by a sort of barrel-organ device, and is said to have performed with taste and to good effect. In these mathematical and musical automata we see the first insidious intrusion of mechanicism into areas that formerly had seemed typical of the rationality distinguishing man from the beast-machine. Consequently, at this moment in time, just before Descartes, began the reaction against automata and the turning back to that mechanistic philosophy which had been their original inspiration.

Related to water-clocks, and producing an almost independent line of evolution for automata in the Renaissance, was the art of waterworks, a technique in which there was almost legendary proficiency in Roman times. From a pair of beautiful Norman drawings of the waterworks of Canterbury Cathedral and its vicinity,[25] we can surmise that the ancient skill was in the hands of able craftsmen by about 1165 A.D., and thenceforth are found the clepsydras of churches and monasteries, depending on an adequate supply of dripping water. During the Middle Ages there seems to have been some production of hydraulically operated automata; authority is lacking, but it is probably safe to assume that they were close to the Heronic tradition in their basic design.

At the close of the thirteenth century a particularly famous set of such water-toys was built for Duc Philippe, Count of Artois, at his castle of Hesdin.[26] It is described in detail by the Duke of Burgundy in 1432, and one gathers that along with the spouts for wetting fine ladies from below and covering the company with soot and flour, were quite a large number of animated apes covered with real hair and sufficiently complicated to need frequent repair. This "pleasure garden," in all its extravagant bad taste, became

25. Robert Willis, *The Architectural History of the Conventual Buildings of the Monastery of Christ Church in Canterbury* (London, 1869), pp. 174–81.

26. Sherwood, "Magic and Mechanics."

the talk of the civilized world and was probably the ancestor of those famous and somewhat more decorous French and English fountains and waterworks of the late sixteenth and seventeenth centuries, whose elegant automata impressed the public and revived in sensitive philosophers the old urge of mechanicism.

Yet another line of development deserves consideration, though it does not directly relate to automata; that is the use of optical tricks to produce apparently magical effects. There is some inkling of this in the writings on optics of Classical antiquity, but plainer mention is made by the medieval Polish scholar Witello. During the Renaissance these optical illusions became quite a popular hobby among the exponents of natural magic and the perpetrators of mechanical trickery.

As a link between the Middle Ages and the Renaissance there is, as in every aspect of the history of technology, the figure of Leonardo da Vinci. His mechanical prowess in automata extended to the illustration of Heronic hodometers and a planetary clock mechanism very like that of de Dondi, both making use of gears.[27] His work on flying machines is well known, but in the present context it may be refreshing to regard it, not as a means for man to fly, but as the perfection of a simulacrum for the mechanism of a bird. He is also reputed to have made at least one conventional automaton, a mechanical lion which paid homage to Louis XII on his entry into Milan by baring its brazen chest to reveal a painted armorial shield of the sovereign.

In the Scientific Revolution of the sixteenth and seventeenth centuries, the dominant influences were the craft

---

27. Derek J. de Solla Price, "Leonardo da Vinci and the Clock of Giovanni de Dondi," *Antiquarian Horology*, vol. 2, no. 7 (June 1958), p. 127. See also, letter from H. Alan Lloyd following *Antiquarian Horology*, vol. 2, no. 10 (March 1959), p. 199.

tradition and the printed book. Both played crucial roles in raising automata, astronomical and biological, to a new height of excellence. In the craft centers, particularly those of the central city-states of Nuremberg and Augsburg, there grew up the first fine workshops of skilled clock and instrument-makers. According to our interpretation of the history of automata, it is no accident that these cities and the whole Black Forest area have been regarded until today as the chief centers for the manufacture of both clocks and dolls. It is equally telling that the product particular to them is the cuckoo-clock, a debased descendent in the great tradition of the Tower of Winds and the Clock of Strassbourg, but one in which is still seen that highly significant liaison between the cosmic clock and the biological artifact.

At Augsburg and Nuremberg during the sixteenth century the masterpieces of the clockmakers were usually extremely elaborate automaton clocks, in which tradition are the brothers Habrecht, makers of the second Strassbourg clock in 1540–74. From about 1550 there are preserved the first of the new series of automated manikins in which the mechanism is considerably advanced beyond the old Heronic devices.[28] For the first time, wheelwork is used instead of levers, gears instead of strings, organ-barrel programming instead of sequential delay devised hydraulically. The skill was so well known that Melancthon wrote to Schöner in 1551, on the publication of the *Tabulae Resolutae,* "Let others admire the wooden doves and other automata, these tables are much more worthy of (true scientific) admiration."

At this stage, half a century before the birth of Descartes, other technologies began to influence the automaton-maker, and his reaction to these in turn affects strongly quite different branches of the sciences, as well as technol-

28. Ernst von Bassermann-Jordan, *Alte Uhren und Ihre Meister* (Leipzig, 1926).

ogy and philosophy. One felicitous example is the use of
the armorer's craft by Ambroise Paré (ca. 1560) for his
design of artificial limbs—partial automata to complete a
man who had become deficient. Then again, the draining
of the Low Countries and English Fens aroused new interest in the hydraulics of pumping engines, and out of
urban development came new ideas in massive waterworks
and portable engines for fire pumps. All these increased the
technical skill of those who would devise the fountains and
automata that were to be the wonder of St. Germaine-en-
Laye, Versailles, and other places.

As for the influence of the printed book we may note
that, although Vitruvius's *De Architectura* appeared in an
incunable edition (Rome, 1486), the works of Heron had
to wait until 1573 (Latin) and 1589 (Italian). Thus,
although the simple water-clocks and sundials described
by Vitruvius were available throughout the Scientific Revolution, the Heronic corpus did not begin to exercise its
greater influence until the last two decades of the sixteenth
century. By that time the craft tradition was already in full
swing and the Habrecht Clock at Strassbourg had been
completed. So, by the time of Shakespeare, man's ancient
dream of simulating the cosmos, celestial and mundane,
had been vividly recaptured and realized through the fruition of many technological crafts, including that of the
clockmaker, called into being in the first place by this lust
for automata.

The new automata were to capture the imaginations of
the next generation, including Boyle and Digby[29] and
Descartes himself. Their very perfection would lead to the

29. Digby's work is specially interesting as the first complex mechanization of plant physiology and as a clear and stated example of influence
by the machines. Sir Kenelme Digby, *Two Treatises: . . . The Nature of
Bodies; . . . The Nature of Man's Soul; . . . In Way of Discovery of the
Immortality of Reasonable Souls* (London, 1658), pp. 255–59.

next phases: automation of rational thought—a stream that leads from Pascal and Leibnitz[30] through Babbage to the electronic computer; of memory by means of the punched tape, first used in sixteenth-century Augsburg hodometers; and of the cybernetic stuff of responsive action perceived dimly in the Chinese south-pointing chariot, decisively in the thermostatic furnace of Cornelius Drebbel, and more usefully than either in the steam-engine governor of James Watt.

Descartes, at the time when the crucial change of direction was about to be made, was probably one of the first philosophers to sense what its characteristics would be. Long before he published his *Discourse,* and perhaps before he had become interested in theology, he toyed with the notion of constructing a human automaton activated by magnets. One of his correspondents, Poisson, says that in 1619 he planned to build a dancing man, a flying pigeon, and a spaniel that chased a pheasant. Legend has it that he did build a beautiful blonde automaton named Francine, but she was discovered in her packing-case on board ship and dumped over the side by the captain in his horror of apparent witchcraft. There is probably no more truth in these rumors than in similar stories about Albertus Magnus and many others, but it does at least suggest an early fascination with automata. And the mention of magnets further suggests the desire to enlarge their potentialities by the use of forces more potent than the mechanical means of the time, an ambition surely presaging the idea that mechanism, now richer in technique than ever before, could simulate the universe to that deeper level of understanding which was indeed soon to be attained.

Descartes's place in all this, then, is that of one who stands on a height scaled and begins the ascent to the next

30. Note that this line of argument makes it significant that both men were philosophers, mathematicians, and pioneers of calculating machines.

plateau, which is suddenly revealed with greater clarity, though distant still. In many ways it is like the balance between materialism and vitalism that would come with Wöhler's synthesis of urea; and, just as in that case, there is deceptively slow adjustment to it among philosophers and a feeling that no ground has been gained or given by either side. From the Lascaux Caves to the Strassbourg Clock, to electronic and cybernetic brains, the road of evolution has run straight and steady, oddly bordered by the twin causes and effects of mechanistic philosophy and of high technology.

CHAPTER 4

The ✡, ✩, *and* ⬡, *and Other Geometrical and Scientific Talismans and Symbolisms*

THE UNSPEAKABILITY of the title of this piece is an attempt to exemplify its thesis. There exists a type of human mind to which the three symbols in the title speak without the intervention of words and in the absence of direct pictorial representation. Such non-representational iconography, it will be shown, forms a long and honorable figurate tradition. It is a fellow to the more familiar literate tradition, common to many cultures and subjects, and the numerate tradition which stands as a characteristic of the quantitative sciences. It is a vital component of the aesthetics of scientific theories, both ancient and modern, communicating a sense of interrelationships amongst a complex "Gestalt" and embodying the principles and the results of theories based on such relationships.

Curiously enough, the figurate tradition seems never to have been discussed in general, although specific instances abound of descriptions of particular diagrams and their uses for magical or scientific purposes. A great deal of confusion arises from the circumstance that the preservation

and transmission of the tradition has depended upon manuscript scribes and copyists who may have been amply competent in literate qualities but deficient in the numerate, as
historians of astronomical tables know only too well, and
in the figurate, as is also attested by many blanks in texts
where the pictures should be. Even when such diagrams
appear, they are often hopelessly garbled by being misunderstood and left uncorrected, and by being veiled in a
secrecy appropriate to their valuable magical content as an
embodiment of potent theoretical understanding. The consequence is that most understanding has vanished and the
modern scholar is unable to develop a history which is
more than a flat statement of instances of the various diagrams. Even then, they appear to be little more than arbitrary emblems that appear and disappear through the
pages of history—as, for example, the well-known and surprisingly recent history of the six-pointed Star of David as
a symbol of Judaism,[1] the five-pointed figure which attains
significance as the pentacle of witches and the Pentagon
Building in Washington, D.C., and such curious symbolisms as the forms of the alphabet letters.[2]

The fundamental quality of a geometric symbol of this
sort is that it gives at a glance a reminder of a theory whose
very elegance is displayed by the form of the lines. A trivial
example can be found in the famous incident of the discovery of the forgotten tomb of Archimedes by Cicero in
75 B.C. when he was quaestor in Syracuse.[3] The tomb, un-
marked by surviving literate description or name, bore as
legend the simple diagram of a cylinder enveloping a
sphere. As such it was immediately obvious to the educated
discoverer as a depiction of the Archimedean rectification
of the spherical surface—unquestionably the most power-

1. Gershom Scholem, "The Curious History of the Six Pointed Star,"
Commentary, 8 (1949), 243–51.

2. S. Goudsmit, "Symmetry of Symbols," *Nature,* 6 March 1937.

3. Cicero, *Tusculanae Disputationes,* V, 23.

fully elegant product of the methodology of Archimedes, and a precursor of the integral calculus. The whole method, the proof, and the results, were keyed to this non-obvious construction, whereby the sphere, whose surface area was to be found, was encased in a cylindrical surface that just touched it and could be compared with it, infinitesimal element by element. That diagram spoke for the scientific personality and achievement of only one man, Archimedes.

A more modern instance might be seen in a popular book by Nobel Laureate Chen Ning Yang,[4] in which the content and the elegance of symmetry principles in the physics of fundamental particles is conveyed in terms of simple diagrams that "speak louder than words." Even the book jacket is a symbol of this sort; it reproduces one of the cleverest and most mind-bending illustrations by the modern Dutch artist M. C. Escher, showing a tessellated formation of mounted horsemen moving in a contrary direction. My point in citing this example is to explain that it is not only the content of modern theory in fundamental particle physics that requires the use of diagrams that would obviously and trivially show the same symmetry as the theory, and indeed of Nature herself. The diagram goes beyond this in assuming a form of such inner elegance and economy that a few lines or simple forms imply a much greater amount of communication than could otherwise be made. Indeed, it would appear that the amount of symmetry and the ingeniousness of its interrelation is virtually an argument for the assumption that this particular theory or set of theories must be true. They must be true because they are so neat and so cleverly interwoven. We shall maintain, furthermore, that when a scientific theory has been developed on such a basis, the diagram tends to take on a life of its own, not just as a representation of the theory or as

4. Chen Ning Yang, *Elementary Particles, A Short History of Some Discoveries in Atomic Physics* (Princeton, N.J., 1962).

an *aide-mémoire,* but as a magical talisman and an object
of contemplation and speculative philosophy.[5]

What we have here is a historically important principle
of elegance which acts, not just as an aesthetic criterion, but
as a guide to the philosophic truth of scientific theories.
Everyone is familiar with the test of Occam's Razor; all
other things being equal, we should prefer the theory that
is simplest, the one that involves least by way of assumptions
and postulates. Now we have in addition to simplicity a
second proof that, all other things being equal, we shall
prefer the theory which displays most of this elegance, this
interlocking Gestalt which seems to force a feeling of neces-
sity and can apparently, in many cases, only be conveyed in
the figurate mode. There would seem to be many strands
in the history of scientific thought where an obscure but
powerful literate tradition is in fact just such a figurate
mode; the obscurity creeps in only through the difficult
process of attempting to translate (as I do now) from the
figurate to the literate. It is perhaps worth noting that a
similar difficulty seems to attend the translation of numerate
to literate. The main threads of Greek mathematics are
literate, but the Babylonian tradition is almost exclusively
numerate in its very sophisticated armory of higher mathematical astronomy.[6] Whenever historians of mathematics
have sought to explain the ways of thought that seem to
pervade Babylonian methods, they are forced to rely on a
method of communication which is that of the wrong

5. For general history but little by way of rational explanation of derivation see:

    Sir E. A. Wallis Budge, *Amulets and Talismans* (New York, 1961).
    Jean Margues-Rivière, *Amulettes, Talismans et Pantacles* (Paris,
1950).
    Kurt Seligmann, *The History of Magic* (New York, 1948), pp. 154,
194, 296–99, 354, 355.

6. See also Chap. 1.

blood-group. Babylonian astronomers seem to have thought of their theories in purely numerate terms, like a stockbroker knowing the state of the market from the ticker-tape alone, without the intervention of graphical methods or statements in words. It seems very likely that the obscure Pythagorean tradition of pre-Socratic Greek philosophy may in fact be at least partially due to a poor literate translation from the numerate astronomical science of the Babylonian contemporaries.

It is also remarkable that the few Babylonian tablets containing figures seem to bear just the type of diagram we shall discuss, in which the connected polygonal and star-shaped "talismans" play a special rôle.

The ultimate foundation for this entire tradition in East and West seem to be the concepts of an element theory. What is at stake is not the predecessor of our modern chemical elements but rather a theory that relates the various forms of substances to all the forces and changes which may be wrought with them and upon them. Thus, element theory contains the rationale of physics, astrology, and alchemy, not just the nature of substance. In particular it should be noted that the concept of atoms is in a quite separate department in the history of ancient science. It seems to derive rather unexpectedly, not from chemistry or physics at all, but rather from a preoccupation with the discovery of mathematical irrationality. The easy proof that $\sqrt{2}$ could not be expressed "rationally" as a number, p/q, had a disastrous effect upon early logicians who were forced to conclude that the integral numbers caused a certain graininess of the universe and forced the abandonment of such intuitive devices as the use of similar triangles in geometrical argument. The style of Euclid's Elements is not so much a pedagogic device of inexorable logical steps, as a successful hunt for a way round the unfortunate hiatus of the irrationality of the real world and its "mathematical atomicity."

The concepts of elements, then, had nothing to do with atoms or other units of substances which could be mixed and compounded like medicinal or culinary ingredients. The element theory had to contain a rationale of forces or qualities that would change and transform one substance to another. The central concept of the four-element theory, the *tetrasomia,* was that the set of basic modalities of matter were produced by the working of two pairs of qualities that acted, so to speak, at cross-purposes to each other.[7] One pair consisted of the opposed qualities of hotness and coldness, the other of wetness and dryness, each set therefore containing a positive and a negative manifestation of a principle that seemed part of the essential character of all substances and all change.

From this central concept a whole theoretical structure could now be erected. The two pairs cross with each other to form the four possible combinations, the four elements of air, earth, fire, and water, each of these terms being taken with the greatest of generality. Air is the symbol and support for all vapors and volatility, earth for solidity, water for all fluids and liquidity; water and earth are visible substances, air and fire invisible. The four elements are necessarily arranged by the crossed principles into a square in which each side corresponds to one of the four periodic exchanges that together comprise a Platonic cycle;[8] fire condenses into air, air liquifies to water, water solidifies to earth, earth sublimates into fire. In the reverse order, fire condenses to earth, earth dissolves into water, water vaporizes into air, and air becomes rarified into fire again.

This doctrine of Aristotelian elements lends itself very easily and naturally to the geometrical symbolisms of figures

7. Serge Hutin, *A History of Alchemy* (New York, 1962), p. 80.

8. Maurice P. Crosland, *Historical Studies in the Language of Chemistry* (London, 1962), p. 29.

composed of a cross or of crosses within squares, or of squares set diagonally within squares (see Fig. 4.1). The

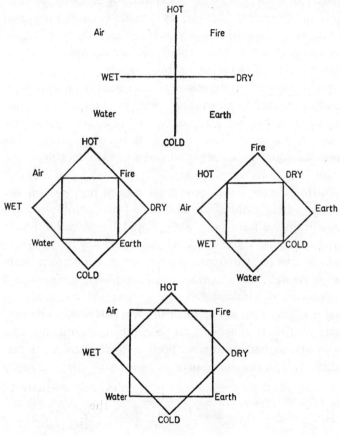

Figure 4.1

antiquity of the figures themselves is indisputably great, but at what period they became associated with element theory is a matter for conjecture. The square figures with diagonals are common decoration found in incised pattern and in tessellations in antiquity. One presumes that the Aristotelian text must have been illustrated originally with some such diagram, and of course innumerable versions exist from the later medieval and Renaissance manuscripts. The whole issue takes on a new significance through the recent identification of the Tower of Winds, built in the Roman Agora of Athens by Andronicus Cyrrhestes ca. 75 B.C., as an architectural exemplification of the octagonal form of the symbolism resulting from the square-set-diagonally-within-a-square form of the element diagram.[9]

In the original archaeological examination it had been determined that this building, perhaps the only surviving classical structure known to have been designed by a mathematician, was an exercise in drawing-board geometry. The orientation along the meridian and a certain determination of the form were essential if the tower was to be used for mounting a wind-vane above, and a set of panels depicting the gods of the eight cardinal winds. Joseph Noble and the present author have been able to make a plausible reconstruction of the water-clock within the tower and to show that the entire structure seems to be intended as a giant cosmic model rather than as a utilitarian combination of a timepiece and wind-vane. There seems good reason to suppose that the form of the building was intended to demonstrate that there must indeed be eight winds and not four or twelve as had otherwise been suggested by rival philosophers.[10] In the same spirit, we suggested that the use of

9. John V. Noble and Derek J. de Solla Price, "The Water Clock in the Tower of Winds," *American Journal of Archaeology,* (1968), 345–55.

10. Note especially that in Vitruvius I, vi, 4, it is stated that Andronicus built the Tower at Athens as an exemplification (*qui etiam exemplum*) of the eight-wind theory or system. Homer and the Bible use the four cardinal winds only, but Hippocrates has a six-wind system and Aristotle uses a zodiacal division into twelve winds. This latter system is exemplified in a stone table of the second to third century A.D., found in 1779 at the foot of the Esquiline Hill and now on the Belvedere Terrace next to the Museo Clementino at the Vatican. On it the twelve winds are named in both Latin and Greek. See James G. Wood, *Theophrastus of Eresus on Winds and on Weather Signs,* (London, 1894), page facing p. 89.

water to turn a sky disc (a star-map in projection) behind an earth grillwork net, probably lit by flames of fire and decorated with playing fountains, was all part of a symbolism of the elements.

Thanks to the publication of an account of the Tower of Winds written by a Turkish traveler in 1668, we are now able to confirm and extend this view on the symbolism and use it as a fixed point in the general history of this figurate mode of thought.[11] The traveler, Evliya Çelebi, though full of fanciful tales and dubious interpretations, indicates quite clearly that the tower also contained some sort of zodiac ceiling, now lost, depicting the twelve constellations and, within them, representations of the planets set in various named signs. The names all agree completely with the standard convention of planetary houses given in Ptolemy's *Tetrabiblos* I.17. The traveler then goes on to speak of a mirror of the world that was once there but now missing, originally set on a pivot—this may well be some misunderstanding of the star-map disc of the water-clock[12]—and adds that there were also 366 talismans, one for each day of the year, and a set of stones such as Yemeni alum and blue vitriol eye-stone, which were related to the black and yellow bile and other humors of the body and were thus of great effect in curing and preventing diseases.

We thus learn that, in addition to the octagonal element symbolism, the tower contained the twelve-sided divisions of the zodiac and a set of associations with planets, humors and lapidary talismans. Some of this theory is well attested by medieval texts; we know, for example, that conventionally in astrology the element of earth was associated with

11. Pierre A. MacKay, *American Journal of Archaeology*, 73 (1969), 468–69.

12. For the tradition of "mirrors of the world" and their identification of star maps, see F. N. Estey, "Charlemagne's Silver Celestial Table," *Speculum*, 18 (1943), 112–17.

melancholic humor, fire with choler, air with blood, and water with phlegm. We know, moreover, that the zodiac cycle began with Aries and springtime and was aligned with air and blood to the south point of the compass and the corresponding wind, as well as to youth. The choice of alignments is not at all arbitrary, but certain key points are obvious choices and, these being made, the rest of the cycle falls into place naturally and uniquely determined so as to form an interlocking set of theories covering virtually all creation and comprehending cosmology, chemistry and physics, meteorology, and medicine. Such was the ambitious burden of the Tower of Winds.

The method of aligning the square and octagonal symmetry of the element theory with the twelve-sided division of the zodiac has a special historical interest. It is not attested in detail by any surviving evidence at the tower nor indeed in any literary text. Nevertheless, the general method by which it must have been achieved has been preserved in the traditional forms of the horoscope diagram, this significance of them never having been noted before. All three early forms of astrological horoscope diagram are formed on the basis of a square intersected either by a cross or by another square placed diagonally over it in a manner very similar to that of the element diagram, and quite compatible with it (see Fig. 4.2) .[13] Once the general principle has been stated, it becomes quite obvious that such a diagram has been used as a basis or rationale for much of the underlying theory of astrological science, and previously obscure alignments and associations may be seen as necessary results of two cycles being aligned from other elements.

13. For the "modern" form see Frederick H. Cramer, *Astrology in Roman Law and Politics,* American Philosophical Society (Philadelphia, 1954), pp. 20, 21; for ancient forms see Cramer, p. 165 and O. Neugebauer and H. B. Van Hoesen, *Greek Horoscopes,* American Philosophical Society (Philadelphia, 1959), p. 156.

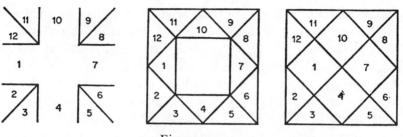

Figure 4.2

It is, I believe, also significant in this figurate scheme that so much of the rest of astrological theory depends on the aspects, particularly those relating one sign of the zodiac to another, where the original text appears to have been illustrated with diagrams (see Fig. 4.3) that serve not so

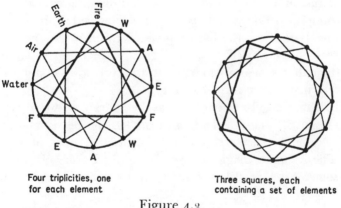

Four triplicities, one
for each element

Three squares, each
containing a set of elements

Figure 4.3

much as illustrations but as figurate theories in this tradition. The figures referred to are those of the triplicities and the squares linking sets of signs distant from each other by a right angle so that they form a square, or by 120 degrees so that they form an equilateral triangle. There are necessarily four of the triangular triplicities, one corresponding to each element,[14] and there are three squares where each square contains a set of elements. Again the alignments come naturally so that Aquarius, for example, must be in the watery triplicity. It seems quite plausible that much of astrological theory may rest on just such a basis of figurate rationality rather than upon empirical or special omen lore. In this sense astrology, quite apart from its utter falsity in the light of modern knowledge, developed on a very rational basis, with a figurate theory and the associated symbolism at its center.

In view of the ingenuity of this matching of the twelve-fold division of the zodiac and horoscope with the fourfold symmetry of the element diagram, it is especially interesting to find that among the relatively few diagrams occurring in the corpus of Old Babylonian mathematical texts we find an entire collection of squares divided in this fashion and accompanied by a text that seems quite enigmatical.[15] Although the text is usually interpreted as pertaining to area calculations for the figures given, I think it may be more reasonably viewed as an exercise in what was peculiarly difficult for the Babylonians, an interpretation of a written text in pictorial form (see Fig. 4.4).

It may also be remarked that the figurate tradition of the cross and square in element theory has also been elaborated to several other well-known and attested magical forms. The standard magic square of the third order clearly has some of the crosslike symmetry of the element diagram and

$$
\begin{array}{ccc}
4 & 9 & 2 \\
3 & 5 & 7 \\
8 & 1 & 6
\end{array}
$$

14. Such a diagram of four triplicities is attested in a Babylonian tablet from Uruk, see F. Thureau-Dangin, *Tablettes D'Uruk,* Musée du Louvre, Department of Oriental Antiquities, VI, (Paris, 1922), plate 26.

15. H. W. F. Saggs, *A Babylonian Geometrical Text,* in *Revue Assyriologique, 54* (1960), 131–46.

Figure 4.4

can be forced into various sorts of agreement with it. With the numbers transposed into alphabetic numerals, it was taken as the source of magical nonsense words in Arabic and in Greek, and it may well be that the famous acrostic word square

$$
\begin{array}{ccccc}
S & A & T & O & R \\
A & R & E & P & O \\
T & E & N & E & T \\
O & P & E & R & A \\
R & O & T & A & S
\end{array}
$$

has been designed with the same symmetry and figurate significance in view.[16] In another variation it may be seen that if one starts from the third-order magic square numbers and draws lines joining the triads of numbers as follows: 1, 2, 3; 4, 5, 6; 7, 8, 9, the resulting figure is the mystic "demon" of the planet Saturn. Very likely many of the other weird signatures and demons have similar origins

16. Charles Douglas Gunn, *The Sator-Arepo Palindrome: A New Inquiry into the Composition of an Ancient Word Square* (Ph.D. diss., Yale University, 1969), p. 235.

in squares of other order. Unfortunately for the four-element theory, there is no possible magic square of the second order in which the totals of rows and columns and diagonals is constant. If there had been, it would doubtless have become a central object in mystic symbolism. The very absence may have, indeed, some indication that the four-element theory could not be a sufficient and complete explanation of all substance and change in nature. It seems, however, more likely that the ingenuity of the explanation was an indication that the theory was on the right track, but in all explanations it became clear that just some little modification would be necessary to make it perfect.

For this reason it seems evident that the four-element theory was followed during antiquity and the middle ages with an elaboration designed to bring it to perfection. I suggest now that there were, in fact, two rather different sorts of attempts to improve the valuable figurative core of the theory and that these resulted in the symbolisms of the pentagram and of the hexagram, respectively.

In the first modification the theory is improved simply by increasing the number of elements from four to five by the addition of a "quintessence." The problem, then, is to determine what in this new scheme can correspond to the neat double duality of principles that was built into the figurate structure of the old Aristotelian theory. By using the complete pentagon, the pentagram taken as their emblem by the Pythagoreans, occurring naturally as a knotted strip, linked to the essential and perfect "fiveness" of the Platonic solids, one could show that the new scheme also had a natural beauty and perfection. If, for example, each side of the pentagon is made to correspond with one of the five elements, the five external and five internal vertices represent all the combinations of elements taken two at a time, and just four such combinations are grouped on each of the lines. Alternatively, the points of the pentagram may be taken to represent elements, and the lines then become relations between them.

In the second modification of the theory, the improvement is obtained not by adding a new element but by adding a third duality to the original two principles. An obvious way of symbolizing all the possible combinations of three intersecting dualities would be by means of three circles in the customary representation of a Boolean diagram of formal logic. It does not seem to have been previously noted that the hexagram, or Star of David or Seal of Solomon, is formally identical with the three-circle diagram. If three alternate vertices are taken to represent the three principles, then the other three vertices represent the combinations of the principles two at a time, and the central hexagonal area represents the combination of all three principles. Furthermore, the sides can also bear interpretation in this way and the whole symbolism can be suitably embroidered and elaborated with the greatest ease.

The possibility that these familiar talismanic diagrams are part of this figurate tradition of an element theory naturally leads one to ask if there are other figures that can be so generated. The figures sought are those formed by the joins of *n* points equally spaced around the circumference of a circle. The system in which each point is linked to the next gives only a regular polygon, an *n*-gon, which appears as a trivial solution. For three points there exists only this solution, the regular equilateral triangle, common enough in the figurate language of mysticism, but not readily bearing any sophisticated interpretation of this sort. For four points the only solution apart from the square is the cross formed by its diagonals and already described as the Aristotelian element diagram that stands near the heart of this tradition.

For five points, the only possibility apart from the pentagon is the pentagram, which has been discussed as a Pythagorean symbol, perhaps illustrating a five-element theory. For six points, again there exists apart from the hexagon, the hexagram which is famous as the Seal of Solomon and Star of David. There exists also the degenerate crosslike diagram formed by the three diametral lines of the hexagon, a sort of set of snowflake axes, but that seems, again, without any significant symbolical properties.

For seven points, apart from the regular convex polygon it is possible to form two distinct types of heptagram; one in which each point is connected to the two vertices distant from it, and one in which each point is connected to the third distant therefrom (see Fig. 4.5). The first of these variants never seems to have been used as a mystical or magical diagram. This is strange, for the second variant is one of the more frequently occurring such instances of the figurate tradition. It is attested on a Babylonian tablet from the Khabaza Collection now in the Philadelphia University Museum,[17] in which it is said to represent the "seven regions" or *heptamychos* of the philosopher Pherecydes of Syros. Astrologically it is very familiar as the heptagram of the weekday gods,[18] in which a diagram containing the planets placed in their astronomical order of distance from the earth is made to yield by jumping three places at a time the order of planets in the days ruled by them in the week. Of course the planets and their gods are also found to be associated with the principal metals, so that this diagram also assumes a special alchemical significance; this figurate tradition, indeed, became central to alchemy since it linked so neatly and temptingly the metal lead designated by the heaviest and most sluggish outermost planet with the goal metal, gold, symbolized by the Sun.

17. See Robert Eisler, *The Royal Art of Astrology* (London, 1946), Plate 16a and p. 273.

18. See, for example, Cramer, *Astrology in Roman Law and Politics*, p. 20.

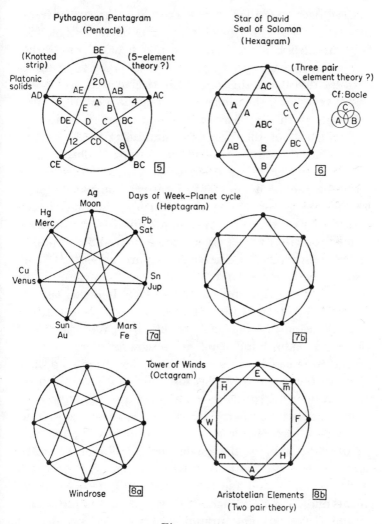

Figure 4.5

For the case of eight points in a circle there exist again two significant forms in addition to the trivial cases of the regular octagon and the star of four crossing diameters. The case in which each point is joined to the next but one has

already been described as that on which the structure of
the Tower of Winds is based: a version of the Aristotelian
two-pair theory of the four elements. It has already been
noted that it has special significance as being compatible
with the division of the zodiac into twelve parts, using one
of the versions of the square horoscope diagram. It has also
been noted that at least one philosopher of antiquity,
Andronicus, took the diagram as indicating a basis for the
eight-wind theory of classical meteorology. This is particularly interesting since the other variant of the eight-
point diagram occurs in many places as a traditional design
for the windrose or compass-card, which is, of course, closely
associated with the winds. I do not think that this association has previously been noted. It may be seen, for example,
on the compass-card of Cecco d'Ascoli, printed in 1521[19] and
also as a basis of the windrose and the grid system of many
portolan charts and other antique maps. It is such an obvious variant and extension of the other eight-point figurate
representation that it seems difficult to separate the traditions and establish independent lineages for them.

Diagrams based on nine, ten, and eleven points do not
seem to occur, probably because they add complications
without increased insight when compared with those already discussed. Similarly, for all greater diagrams we find
no evidence except for what is undoubtedly the most famous tradition of all, the duodecimal division of the zodiac
and the associated astrological theory replete with trines
and sextiles, squares and triplicities, and other such alignments and correspondences. It may well be that just such a
technique of skipping around a circle, well known from
Seleucid astronomical mathematics, may be at the origin of

19. Silvanus P. Thompson, *The Rose of the Winds: The Origin and
Development of the Compass-Card* (Read at the International Historical
Congress, April 1913, from the *Proceedings of the British Academy,* vol. 6
[London 1914]) p. 11.

this entire corpus of figurate methods though, as has been remarked, the evidence concealed by mysticism and bad copying is too difficult to follow at this stage.

The figurate tradition of all these related polygonal diagrams having now been explored, we must turn finally to what appears to be a relatively small collection of other varieties, including some from other cultures. The existence of one series points to the possibilities of others. Is it entirely capricious to see some association between the Yin and Yang diagram symbolizing the paired principles of Chinese elemental philosophy, the three-legged *triskelion,* and the four-legged swastika? Each of these occurs in left- and right-handed varieties, and they can be set in a series as curvilinear partitions of a circle.

Quite different is the case of the mystic Hebrew figure known as the Sefirotic tree of the Cabalists.[20] Although the diagram contains correspondences between the letters of the Hebrew alphabet, the elements, seasons, parts of the body, days of the week, months of the year, etc., it seems evident that the system is based not so much on the shape of the diagram as upon the sequence and significance of the letters of the alphabet; the tradition is indeed much more literate and perhaps numerate than figurate.

Lastly, and most diffidently, I must consider the Chinese tradition. The essentials of the five-element theory are well known and I can add nothing to the historical evidence.[21] For five elements there must be $4 \times 3 \times 2 \times 1 = 24$ different arrangements around a circle or a pentagon, but of these half are mirror images of the rest, and there are therefore no more than 12 basically different arrangements of this sort (not 36, as maintained by Eberhard and followed by Needham, *Science and Civilization in China,* p. 253) . It

20. See Seligmann, *History of Magic,* figs. 155, 156.

21. Joseph Needham, *Science and Civilization in China,* vol. 2 (Cambridge, 1956), especially section 13d, pp. 253 ff.

is interesting that only three of these twelve seem to attain considerable importance as a sequence of physical significance. Perhaps more significant from the figurate point of view is the tradition that comes near to the Aristotelian four-element theory and may well be the origin of the elaboration of this to include a quintessence. In this, the element Earth is placed at the center of the (square) diagram, and the familiar pair of elements, Water and Fire, occupy the north-south axis. On the east-west axis, however, instead of the Air/Earth combination of the Aristotelians is the peculiarly Chinese duality of Wood and Metal. As before, the set of elements is aligned with several other sets of properties and objects. The zodiac is presumed to run from Aries in the east, clockwise via the south; planets and colors and, as a very Chinese touch, tastes are given their alignments (see Fig. 4.6). In quite another version there is a Chinese figurative scheme which seems to be in the same

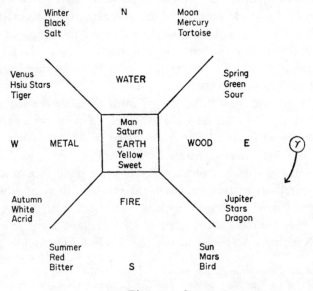

Figure 4.6

tradition, the hexagrams, a set of eight triplets of whole or broken bars—essentially a set of three-place binary numerals. These are associated with the eight compass directions, similar to the Western diagram of the winds associated with the Tower of Andronicus at Athens. In this set, however, we have associated the set of five elements, augmented by other things like mountains and wind, thunder and lightning, and with water occurring as fresh and salt varieties. I think it is likely that the specific association of particular trigrams with their designated elements follows a rational and figurate scheme, probably through a topological correspondence which must exist between the trigrams and the six-pointed Seal of Solomon figure already discussed; each of them is merely a formalized version of a Boolean logic diagram showing the overlapping of three logical classes. It seems very likely, too, that the alchemical symbols of both East and West may draw quite heavily on this sort of figurate tradition, the relevant portion of the element diagram standing as a symbol for a particular element or combination of them.

CHAPTER 5

Renaissance Roots of Yankee Ingenuity

EINSTEIN had just one marvelously simple theory—concerning the motivation of scientific research. He used to say, "You can't scratch if you don't itch." Judged both by the amount of scratching done by modern historians of science and that done by the seventeenth-century participants in the process, the Scientific Revolution was by far the biggest itch there has ever been in the hide of our civilization.

Never has any revolution been so well planned and foreseen, so effective in its execution, and so radical in the way in which it changed society. Seldom has any historical process seemed so disarmingly clear in its structure and essentials. From the standpoint of those to whom it is self-evident that a certain methodical sequence is inevitable in science, the role of the historian was merely to confirm the facts and supply names and dates.

Thus, Francis Bacon plotted the revolution and codified the scientific method. Galileo upset the scholastic philosophy by erecting the art of systematic experiment. Newton carried both processes to new heights. He wielded his powerful mathematical techniques to such advantage that terrestrial and celestial mechanics were united and astronomy

could at last answer "why?" instead of only "how?" Though
the names might be multiplied, the principles must remain
constant. Science is obviously a matter of geniuses making
a sequence of mighty discoveries.

Unfortunately, when examined in detail, this story turns
out to be misleading and unsatisfying. Not only does it contain a certain amount of trivial error, but also it evaporates
with disconcerting speed if one seeks any sort of reason for
the existence of a Galileo at one particular time and a Newton at another. As each man is related to his scientific
environment, one finds that, although Bacon was taken as
an emblem on the shield of the Royal Society, he was, in
truth, only the most publicized preacher of a method that
had been growing for decades before him. Galileo, rather
than breaking with the past, is perhaps more accurately to
be regarded as rounding out a process of refinement of
mechanics which had matured during the Middle Ages but
had lain dormant for a century.[1] The famous story of the
Tower of Pisa is perhaps false in all essentials, and his main
work dependent more upon thought experiments than any
real trials with apparatus. In the same manner, we find
Newton tightly linked to the running battles that had been
fought in mathematical astronomy ever since Kepler had
cracked the Ptolemaic theory a century before. The picturesque story of the apple, though probably more true
than not, is a dangerous myth if it leads us to think that
Newton's triumph arose solely from an inspired speculative
revelation. Although the "Eureka Syndrome" is phenom-

1. The pre-Galilean story of mechanics has now been documented and
edited with the greatest scholarly care by Marshall Clagett in *The Science
of Mechanics in the Middle Ages* (Madison, 1959), and there is no longer
any excuse of the unavailability of source materials in this field. A very
critical examination of the Tower of Pisa incident has been made by Lane
Cooper, *Aristotle, Galileo and the Tower of Pisa* (London, 1935), but doubtless the picturesque story will linger on as part of the modern mythology.

enally common in science, Newton's theories arose from a long-standing itch that he scratched for many years.[2]

The Scientific Revolution did not arise suddenly and out of nowhere through some mysterious generation of a set of unprecedented geniuses at that time and at no other. It is a product of certain demonstrable forces and ancestry, and in seeking a strategic line through this history we must first exorcise from our mythology all the great men. Any attempt to do this immediately raises the hackles of all good scientists, and it is rather instructive to stop for a moment and recognize the seat of those emotions connected with anything that seems to be a denigration or belittling of the heroes of science.

In ordinary history the process has long been familiar. As Alexis de Tocqueville remarked, in an age of aristocracy the attention of the historian is focused upon the heroes, kings and queens, and great leaders, but in an age of democracy the tendency is to consider the general process, a trend clearly apparent today in the work of the social historian. Now science seems so essentially a democratic process that perhaps the judgment of Tocqueville holds here too, though the instinctive reaction of the scientist against such treatment seems to be stronger.

The psychology of the reaction is most interesting. Science seems tied to its heroes more closely than any other branch of learning. It is the one study that contains the entirety of its successful past embedded in its current state; Boyle's Law is alive today as the Battle of Waterloo is not. Because of this, the history of science is capable of much deeper and more logical seeking toward a general history

2. Several entertaining examples of the Eureka Syndrome at work have been collected and discussed by R. Taton, *Reason and Chance in Scientific Discovery* (New York, 1957), translated from the French by A. J. Pomerans. Taton finds that the *Geistesblitz* generally appears not during periods of assiduous work but rather during those of rest and relaxation.

than most other branches of history. Owing to the perpetual immanence of its past, science is conceived in the public imagination as something dead and cold and logical; dead scientists are honored, but the living ones are felt to be apart from common humanity—though of course in private experience we may find individual scientists to be most delightful, lively, and cultivated persons.[3]

Then again, the motivation for research may be an intellectual itch—indeed, the purpose of education has been defined as the business of making people uncomfortable, making them itch—but a deeper and more specific urge may have made these persons into scientists. By far the most common inner reason is that as youngsters they have wanted to be a Mr. Boyle of the Law. They seek an immortal brainchild in order to perpetuate themselves. In an age of teamwork amongst scientists, of little men working on big machines, this hallowed form of eponymic immortality is becoming insecure, and the image of really great men and their theories has become more precious. If, however, this is becoming a problem, there is surely all the more reason to examine the process that made it possible, during the Scientific Revolution, for men to fashion bricks of science inscribed with their own names and build up, faster than ever before, an imposing edifice and superstructure of theory and experiment.

Why did the Scientific Revolution happen when it did? Quite certainly it is a product, in some way, of the Renaissance. It is not, however, the momentous rebirth of classical aesthetic forms that one knows so well from the visual arts,

3. The public image of science has been devastatingly exhibited in a pilot study which is now so often cited as to be considered a *locus classicus*, Margaret Mead and Rhoda Métraux, "Image of the Scientist among High-School Students," *Science, 126* (August 30, 1957), 384–90. For further variations on this theme and an analysis of its consequences, see Gerald Holton, "The False Images of Science," *The Saturday Evening Post* (January 9, 1960), pp. 18 ff.

or, indeed, the ferment of recaptured literary styles and the evolution of classical philology and scholarship that followed them, that must concern us; it is rather the rebirth of the scientific knowledge of antiquity. And in this the thunder of the Renaissance had been stolen by the other renaissance of the Middle Ages.

The flickering torch of late Roman scientific learning had been passed through Byzantium and several other cultures of which we have only monumental ignorance and was breathed into active life again by the world of Islam as soon as it had settled from its initial evangelism under Mohammet. From the eighth century through the thirteenth, the fire burned bright, and much was added in all fields of learning. Then in the twelfth century, principally in the linguistic and cultural melting pots of Sicily, Toledo, and a few other places in Moorish Spain, there came the Age of the Great Translators. They took the corpus of classical learning and its Islamic overlays and translated the greater part of it through a multiplicity of languages into Latin. In this form it became known in European universities during the twelfth and thirteenth centuries and led immediately to furious activity and to more and highly original work.[4]

Thus, there were periods in the Islamic and European Middle Ages that produced wonderful new work, and effected much more than a simple transmission of texts from Greek to Arab and from Arab to Schoolman. Nevertheless, since transmission did occur, all the great ancient works of learning were available in the West by about 1300. By that time there was little left for any further renaissance to ac-

---

4. The best general story of the great translations from Arabic into Latin is told in Charles H. Haskins, *Studies in the History of Medieval Science* (Cambridge, Mass., 1924). More popular versions are available in the same author's *The Renaissance of the 12th Century* (Meridian Pocket Books, New York, 1957), and in A. C. Crombie, *Medieval and Early Modern Science* (New York, 1959), especially *1*, Ch. 2.

complish in the world of scientific learning. The main task
had been completed. One might perhaps have expected a
steady growth from here on, with only the added and later
impetus of a recovery of classical aesthetics.

It did not happen that way. In the realm of science, as
indeed in its economic life, the Middle Ages began to die
on its feet by the end of the fourteenth century. One sees
a very marked decline in the fire and originality as well
as in the number of writers on learned matters. The Merton
College school of great astronomers and mechanicians collapsed by about 1390. The University of Paris declined
after the time of Nicole Oresme. There was a period extending from about 1400 until 1460 when science was as
dead as it has ever been. Probably, like all Dark Ages the
phenomenon is partly attributable simply to the ignorance
of the modern surveyor of the scene, but it seems plain that
some decline had set in.

There can be little doubt that what rescued scientific
learning from oblivion was the invention of printing and
its rapid growth in Europe from 1470 onwards. In itself,
this invention occupies an important place in the history of
technology and is clearly associated with the ever growing
ranks of the high technologists during the Middle Ages.
It is, however, because the effect of the invention was so
cataclysmic—in a good sense—that we must stop and examine the process. It all worked rather like the process we
have seen unleashed within the last two decades, the revolution in publishing caused by the paperback book.

The first stage was a ransacking of the entire available
corpus of the classics for republication; there were even
presses, such as that of Regiomontanus in Nuremberg or
Ratdolt in Venice, that specialized in science, like the modern Dover reprints. The second stage was an active hunting
of new manuscripts and a press-ganging of all available
contemporary writers. Like the rise of the paperbacks, the

original mushrooming of printing changed the book habits and the scholarly machinery of the nations. At first, printing merely relieved the pressure on the copyist manuscript scribes as the paperback has relieved the pressure on the secondhand bookseller.

By about 1500 the age of the incunabulum was over, and the printed book had become a quite new force.[5] The momentous effect, of course, was that the world of learning, hitherto the domain of a tiny privileged elite, was suddenly made much more accessible to the common man. In religion it is clearly this process, more than anything else, that lent strength to the Reformation. As one would expect, in lands where the Reformation was strong, the rapid mobilization of the new learning was also strong. It was the Germanic region of Luther rather than Catholic Italy that saw the revival of astronomy by Regiomontanus, Kepler, and Copernicus. At the beginning it seems as though this positive force, rather than any antagonism of religion toward science which might have grown up later, provides a signpost.

To take stock of the sixteenth-century changes that were promoted by this flood of books, we must review the raw material that was available at the beginning of the century. The two legs of science were its mathematical physics on one side and its high technology of scientific instruments on the other; carried along by the momentum of these parts were assorted pieces of the chemical and biological arts and sundry theories and mechanic skills that had not yet been incorporated into the anatomy of the legs. The first effect of the printed word was to communicate both mathematical

5. The prime source for bibliographical information on scientific books before 1500 is A. C. Klebs, *Incunabula scientifica et medica, Osiris, 4* (Bruges, 1938). An analysis of this by George Sarton, "The Scientific Literature Transmitted through the Incunabula," appeared in *Osiris, 5* (1938), 41–247, later summarized by the same author in *Appreciation of Ancient and Medieval Science during the Renaissance* (Philadelphia, 1955).

methods and mechanical devices to a far larger audience.

In its reaction, science behaved exactly like an atomic explosion. It had done this before. The unification of Greek and Babylonian astronomy may be compared to a fusion bomb in which the parts conflated with the release of much surplus energy. The business of the books worked much more like a fission bomb in which critical mass had suddenly been attained and a chain reaction produced. At this time science became cumulative in a way it had not before. In previous ages each man had made his contribution based on seemingly age-old wisdom which he had learned in his early training. Now the pace became faster, so that a person had to read quite new books, and even keep up with the work of his contemporaries, in order to advance. At this point, however, the device of the scientific paper had not yet been invented, and men did not publish until they thought they had mastered completely some whole department of science and could produce a definitive book. The next stage—the coming of the scientific academy and its learned journals—did not happen for another century and a half, in the middle of the seventeenth century, when the Scientific Revolution was already well under way.

In that period of 150 years we have our giants, such as Bacon and Galileo, Gilbert, Harvey, and the young Newton. These are, however, only those who wrote the successful definitive books. To be understood they must be seen against the background of the extras on the stage of science: those who were reached by their books and were moved to make and use instruments but did not themselves make individual contributions for which they are remembered. Such people are neglected by historians, partly deliberately as minor irrelevancies; partly, however, through the intrinsic difficulty of finding out anything about them. Their writings, such as they published, are by definition rare and second-rate. The greater number of them were practical teachers or working artisans. The former leave traces only in fugitive examples of syllabuses and students' notes. The latter may often be known solely from their rarely preserved instruments and artifacts.

The arduous task of assembling data for the early mass movement in science has, in spite of all difficulties, been now accomplished for many countries and areas.[6] The most obvious, yet remarkable, finding of this study is the enormous number of minor characters of science that worked even before the days of the Royal Society; perhaps one must admit also that these little men may well have had a bigger effect in total than any one of the giants of genius. Certainly they cannot now be neglected as part of the story.

The earliest band of scientific practitioners in England were the surveyors, who found increased employment in the redistribution of lands consequent upon the dissolution of the monasteries. There were also the early teachers of arithmetic for mercantile use, the teachers of navigation, and the makers of magnetic compasses. Most of the earliest instrument-makers were immigrants, many of them refugees from religious struggles on the continent. The greatest fillip to the artisans came when Elizabeth decided not to rely on foreign powers for her brass cannon and founded at home the Mines Royal and Battery Company. This made available for the first time in England a source of good brass plate. There were all sorts of unexpected repercussions of this. For one thing, the church brasses cease to be imported and become more numerous as a home product; for another, this marks the beginning of a large-scale in-

6. The classical study of the mass movement in science is E. G. R. Taylor, *The Mathematical Practitioners of Tudor and Stuart England* (Cambridge, 1954). For France, there is Maurice Daumas, *Les Instruments Scientifiques aux XVIIᵉ et XVIIIᵉ Siècles* (Paris, 1953). For Germany and several other countries around it, Ernst Zinner, *Astronomische Instrumente des 11. bis 18. Jahrhunderts* (Munich, 1956). For the Low Countries, Maria Rooseboom, *Bijdrage tot de Geschiedenis der Instrumentmakerskunst in de noordelijke Nederlanden* (Leiden, 1950).

dustry in the manufacture of all the astronomical and other instruments that are best made from brass plate. As a matter of fact, one of the chief men of the Mines Royal, Humphrey Cole, a northcountryman who had earlier worked in the Royal Mint, became the first great instrument-maker of England and produced many of the navigational aids for the famous voyages of Elizabethan discovery.[7]

Another man of the times was Thomas Lambritt, alias Geminus, a refugee from Lixhe, near Liége, who engraved masterful astrolabes and other devices. He is known also as the engraver of the wonderful anatomical plates which illustrate Vesalius, and this underlines the very close connection which existed between the arts of scientific instruments and the process of copper engraving that was so important in the sixteenth-century book trade.

From such small beginnings the labor force of practitioners grew, multiplying as each master trained some three or four successful apprentices who later became independent. E. G. R. Taylor lists more than a hundred known names before 1600, and nearly 250 by the middle of the seventeenth century, virtually all of them in London. In 1650, before there was any formal organization of the Royal Society, there must have been more than a hundred such artisans and practitioners gathered in dozens of independent establishments all over central London—a very sizable activity and industry, even for so large a town, in this period.

In fact the very inception of the Royal Society may be rather directly attributable to the practitioners. Before the days of their Royal Charter, the amateurs met as a club, later called the "Invisible College." In the beginning it was entirely informal and centered not only on the chambers of its chief participants but also upon the shops of the instru-

7. For the life and works of Humphrey Cole, see R. T. Gunther, "The Great Astrolabe and Other Scientific Instruments of Humphrey Cole," *Archaeologia*, 76 (1926–27), 273–317.

ment-makers and the taverns (later coffeehouses) they frequented and used as a sort of general post office. Eventually, when the club met more regularly, it seems to have been called together by Elias Allen, chief of the instrument-makers. He was an apprentice, several times removed, of Humphrey Cole, and, acting as a sort of union organizer, he mobilized the instrument-makers and led them in a block to join the guild of the Clockmakers' Company.

It must be insisted that although these men had been called to their trade by the usefulness of the things they produced and taught, this usefulness was not sufficient for their support. They were, for the most part, powerless dupes of the process of democratization of science by the flood of books and the spread of mechanical ingenuity. It was they who had to seek out the scientists and the amateurs of science and make them feel it was a smart and cultivated thing to buy a microscope or a slide rule.

One has only to look at the entries in the diary of Samuel Pepys to realize how proud he was to buy a calculating rule and optical instruments and be taught the secret delights of their use. Pepys was indeed a very special amateur. Not only did he become Secretary of the Navy but he rose also to the presidency of the Royal Society. It was he, indeed, who affixed the *imprimatur* of that august body, on the *Principia Mathematica* of Isaac Newton.

For the early practitioners it was uphill work in salesmanship, though, for the record shows that most of them lived in acute poverty and died of starvation. Even the first paid scientist, Robert Hooke, who was employed by the Royal Society to "furnish the society every day they met with three or four considerable experiments," had impossible difficulty in getting money for his work. At one time he was paid off with copies of a book on fishes published by the Society but not sold very widely—poor recompense for a production line of several new discoveries a week!

It is against such a backdrop of minor actors, men who earned a precarious living from practical pursuits or from teaching such practice, that one must view the activities of their clientele of scientific amateurs and the few genius members of that clientele whose names have become household words of science. Galileo making his telescope and clock and Newton experimenting with a prism and making the first reflecting telescope are contained well within the province of the practitioners. Only otherwise, when they write their monumental books, do they rise above it.

We must now go further into the character and consequences of this mass movement in science. One effect, the clearest, is the rapid organization, almost simultaneously in several European countries, of formal academies of sciences where the now numerous band of amateurs and even professionals could meet, exchange views, and share the services of an "operator" and the expensive instruments and collections. From this, in turn, arose the very conscious invention of the scientific paper as a device for communicating and preserving the knowledge that was now accruing at a rate faster than could be assimilated into definitive books.

Another effect, not nearly so clear but just as vital to the life of science as the learned journal, was the way in which the practitioner movement led to the establishment of experimental science. The public image of the modern scientist as a man-in-a-white-coat-in-a-tiled-laboratory is so strong and pervasive that one has difficulty in regarding it as perhaps but a recent pimple on the body politic of science. The public laboratory as an academic or industrial institution is barely more than a century old. It arose first in chemistry about 1840; perhaps Liebig's laboratory in Giessen is the best known of the pioneers. In the 1870's it entered physics—the Cavendish Laboratory (Cambridge, England), opened in 1874, was the first building architecturally designed as a place in which to work with physical apparatus. The laboratory at Oxford had been modeled after the kitchens at Glastonbury Abbey, a large place where one could cook chemicals.

In the midst of today's urgent activity in the provision of laboratories for high schools, it is sobering to reflect on the rapidly changing—nay, ephemeral—condition of the scientific laboratory. Less than a hundred years ago the laboratory was a place where the use of rather complex and expensive instruments could be learned and shared; then it became a storehouse of unit devices and apparatus that could be connected in various ways and improvised with sealing wax and string to do all the new things demanded by the explosively accelerating research front; then, gradually, certain pieces of apparatus got larger and larger. Giant electrical machines were already produced in the eighteenth century. In our own times, the first miniscule cyclotrons built by E. O. Lawrence in 1929 rapidly grew into operations so costly that their administrators speak of "megabucks." The current pattern is clearly exemplified by the giant machine, envisioned by scientists, built by engineers as a piece of apparatus that is an institute in its own right, and staffed by teams of quasi-anonymous slave-laboring Ph.D. candidates.

It would be rash to suggest that the old style of physics laboratory is doomed, rasher to say that a similar thing must happen eventually in the later-developing subjects of chemistry and biology. Yet clearly we have here a state of considerable flux, and the stretch of memory of living men is not to be taken as an infallible guidepost to the future. The laboratory, as we see it now, is not nearly so historically fundamental in the life of science as is the general use of observation or the quite basic mathematico-logical formulation of science.

Returning now to the wider historical problem, we can

see the first, tentative nineteenth-century public laboratories as a logical continuation of the old, private process. Galileo and Tycho Brahe had employed their own workmen and bought from the ingenious artificers. Pepys had kept his calculating rule and perspective glass on the shelves with his books. Newton had his prism and telescope in his study. Even in the early nineteenth century, only a man in a special position, like Michael Faraday at the Royal Institution, could enjoy the purchase of the increasing range, and afford the rising expense, of instruments. Eventually a point was reached where one man could no longer work privately. If he was a professor, he had the advantage of being able to use the more promising students to stir his calorimeters. In the universities too, even at an earlier stage, it had become common to acquire apparatus for the purely pedagogic purpose of exhibiting impressive experiments in the courses on natural philosophy.

Thus it is that science strode on its two legs through the Scientific Revolution and toward the Industrial Revolution. When experimental methods and instruments had reached sufficient maturity, they began to feed back upon the body of science and destroy the former lag between the development of new tools and their application. This can first be seen in the seventeenth century, when the Royal Society, under the slogans of Baconian New Philosophy, is really self-conscious about applying its freshly won knowledge to the betterment of mankind. In France, this governed the process of the Enlightenment, and the whole philosophical teaching of Diderot in the *Grande Encyclopédie* is carried within the format of a scientific elucidation of the trades and industrial crafts of the people.

In the nineteenth century, the modern process of industry was particularly striking in the development of electrical science and of the industry based upon it—the first great technology to arise directly out of a new branch of science.

Each new conjuring trick produced by ingenious apparatus added something to the science; each new advance in the science produced a host of further tricks and also the by-product (which eventually becomes a main product) of new apparatus and new machines. Elsewhere in the regions of science one finds old, low technologies suddenly becoming complex and a prolific growth of new machines and methods for making them. At the heart of the matter always, though, there are the ingenious mechanicians, the unschooled amateur scientists and artisans, the cultivated patrons of these workers, and all the other little men of science, bound together by the mass movement. Only in more recent times, when the scholarly elite is no longer separate and the big men are only little men magnified from the common stock, do we realize that this aspect of the Industrial Revolution has become sensibly complete.

When I was first brought face to face, as a visiting foreigner, with the problems of the history of science in America, I was deeply puzzled. Here we have the phenomenon of a country pre-eminent in high industrial technology and in all the pure science that goes with it. How did this come about? The scientific achievements of colonial times —even indeed the sum total state of science in this country up to about a hundred years ago—seems to have a surprisingly small absolute value, even for a land whose chief worries were in other regions of human endeavor. What is most perplexing, however, is to consider this low state of science relative to the expansion which did in fact take place on so apparently unpromising a basis.

It seems almost ludicrous, in terms of historical perspective, to say that a Franklin and a Priestley, even with a dozen others, and aided and abetted by such latecomers as Willard Gibbs and his colleagues, were wholly responsible for the local climate of science. In more recent times the contact with Europe has been close and the traffic of scholars and refugees crowded, but this cannot explain completely the earlier period or the specifically national characteristics, such as they are. It seems equally irregular to attribute the Industrial Revolution in this country to the capricious appearance of a band of natural inventors such as Eli Whitney, Edison, and Ford. Even though we make suitable national obeisance at the shrines of these mechanics and scientists, as indeed do the Poles for Copernicus, the Russians for Popov and Tsiolkovsky (the rocket inventor), the French for the Curies, and the English for Newton, it is a false honor and a disservice to their names to ignore their scientific contexts.

As we note in connection with the Scientific Revolution in Europe, the giants are better seen against the backdrop of the little men. I suggest it might be profitable, now that we have reached this recent understanding of the European process, to apply the same considerations in the field of American studies and seek the extent and influence of some practitioner movement here too. Of course, such transference may not be carried too far or pursued without due caution. For one thing, the movements are not by any means contemporary; such events in America seem to be about a century later than comparable happenings in Europe. For another consideration, there is the curious difficulty that although we may seek and find national differences in science, there is a wider sense in which the pursuit of science and its corpus of knowledge are overwhelmingly supranational.

It takes a great deal of wartime secrecy or geographical isolation to make the local state of science in any place fall above or below that of the universal body of knowledge, even for a very limited period of time. One book can leak the knowledge of centuries; one man of similar training faced with the same problems can duplicate unwittingly another's research and experience the embarrassment of coincidental publication or application. In this respect it is precisely the giants in science, the men of genius, who provide the least clues to an understanding of any domestic issues. Franklin, Newton, Galileo are only in a very limited sense national heroes if one considers their scientific work. They belong to the world.

Thus, for the history of science, it may be actually more convenient to regard Franklin and Priestley, perhaps also Willard Gibbs, as European scientists who happened to be on the other side of the Atlantic. They happened to have been here, but perhaps it is not too bold an exaggeration to suggest that it might have had little effect on the destinies of American science if this little band of geniuses had been much more numerous or much less. Certainly I feel grave difficulty in proceeding from a Franklin and a Priestley toward an understanding of some special brew of men that produced scientific America.

How do we fare, then, if we look for some analogue of the practitioner movement here? I feel we fare exceedingly well. Consider the elements that are available for inspection. Everywhere in colonial history one meets the enthusiastic amateurs of science, eager for experimental science and the practical application of instruments to surveying and navigation and other arts. They are not scientific geniuses, but they often do good, solid bits of work. They flourish in groups, enjoying stimulating philosophical conversations, and they patronize and support the efforts of the ingenious artisans, mechanicians, and other practitioners. They are fully comparable to Samuel Pepys and his cronies of the early Royal Society.

To name but a few of them, one might take the incomparable Thomas Jefferson, John Winthrop, and the numerous men of good will whose labors founded and promoted the early colleges of America. Mentioning the colleges, one must surely include many of the men who taught the sci-

ences at these places. Some few might be counted as trans-
atlantic European scientists and professors, but the major-
ity seem to be much more of the breed exemplified by
Robert Hooke, the ingenious artisan who demonstrated
experiments, and who, though reasonably educated, had no
special training in science other than that acquired by ap-
prenticeship and application to the art.

When we come to the avowed practitioners of science, the
volume of evidence seems overwhelming. In America as
in England, the early instruments and their makers were
imported, but this movement later declined with the domestic development of men who were able to do the job.
The difference here is only that America was a much bigger country, and the movement to the frontier left gaps
that needed further replenishment by importation.

There are some curious features of the importation process. Many colleges had friends or formal agents who
brought scientific instruments for them from Europe. At
one time, the wild sabbatical to London or Paris to buy
apparatus and books was the best chance of travel for a
professor—a late eighteenth- and early nineteenth-century
equivalent of a Fulbright Fellowship. This became particularly widespread just after the expansion of colleges
and laboratories resulting from the Land Grant Act of 1862
and was perhaps one of the strongest links with European
science in that period. Many of these instruments are still
extant—some of them of the greatest interest and beautiful
as examples of the finest functional craftsmanship.[8] The

8. The most complete treatment of any American collection is to be
found in I. Bernard Cohen, *Some Early Tools of American Science* (Cambridge, Mass., 1950). Another, more recent *catalogue raisonné* is Leland
A. Brown, *Early Philosophical Apparatus at Transylvania College* (Lexing-
ton, Ky., 1959). The only work that attempts to collect general details of
American instrument-makers and other practitioners (mainly in chemistry)
is Ernest Child, *The Tools of the Chemist* (New York, 1940). A card index,
as yet unpublished, of all American instrument-makers known through city

returning men themselves were objects of curiosity, and many joined in the typical practitioner activity of giving popular lectures on science, illustrated by demonstration experiments. Some of them, as well as some native autodidacts, toured the country as itinerant lecturers—a philosophical analogue to the gospel preachers.

Well established and flourishing in the big cities there were all the familiar facets of practitioner activity. Men like John Ellicott were making surveying and astronomical instruments recognizably based on European prototypes but constructed by eye with methods improvised and alien to the old tradition. David Rittenhouse built most complicated orreries that attracted great attention, just as had the comparable instruments in Europe, though the American worked from first principles, needing only the stimulus diffusion that indicated the machine could be made. Consider in this respect such an example as the Folger family of Nantucket Island, a practitioner clan to which belongs Ben Franklin and also the gadgeteer and congressman Walter Folger, who built in 1785 what is perhaps still the most complex astronomical clock in America, yet preserved and ticking away in the Historical Museum in Nantucket.[9]

Then, again, there is the occasional giant among practitioners—for example, Nathaniel Bowditch, who restored and vastly improved the whole science of navigation and incidentally translated the works of Laplace from the French in four tremendous tomes that add three times its weight in commentary to the original text. Hanging to his apron

------

directories and through signatures on extant instruments is maintained at the Division of Science and Technology, U.S. National Museum (Smithsonian Institution), Washington, D.C.

9. The story of Walter Folger and his masterpiece has been well told by Will Gardner, *The Clock that Talks and What it Tells* (Whaling Museum Publications, Nantucket, 1954). For that other masterpiece of Yankee clockmaking, see Howard C. Rice, Jr., *The Rittenhouse Orrery* (Princeton, 1954).

strings were a horde of makers of navigating instruments, lesser teachers of navigation, and masters of that art who flourished in the great ports and gradually extended their domain to such other activities as surveying and chart-making.

In surveying itself there is the immortal pair, Charles Mason and Jeremiah Dixon, generally remembered only through the politics of the line they surveyed. Yet these imported English specialists were accomplished surveyors whose scientific worth in technical astronomy was of the highest caliber and who did much to open the country culturally as well as delineate it geographically. One has only to observe the unusual occurrence of straight lines all over the map of the United States, in roads as well as in state boundaries, to realize that here is a country more heavily indebted than any other to the work of the surveyor. It is surely not unreasonable to suggest that this work must have had influences that run deeper than the sheer achievement of criss-cross patterns on the map.

For scientific instrument-makers, one need only examine the nineteenth-century city directories of Boston, Philadelphia, and New York to find hundreds of names of craftsmen and firms. It is, to be sure, an antiquarian research, for one does not expect to find great discoveries coming from these people. But, just as in Europe, it is a populous trade, influential in the growth of science and highly effective in spreading and intensifying the itch for ingenious instruments and devices. It is by these men that the basic skills of the Industrial Revolution were populated, and it is to them that we must ultimately attribute the phenomenon of the giants who had their brains on the tips of their fingers and in their hands. In such a matrix it is not so anomalous to find an Edison, a Ford, or a Samuel Morse.

Further, there are at least two entire special fields of practical activity in which the country enjoyed peculiar incentives by virtue of its geographical position. One of these was astronomy, where the distance from Europe made possible a series of important observations, such as the sighting of comets and the visibility of eclipses of the sun and transits of Venus. From this a great boost was given to the building of telescopes and observatories. By the mid-nineteenth century, such men as Henry Fitz, of New York, and Alvan Clark, of Boston, had outdistanced their European counterparts. Thus began a short but ecstatic period when observatories broke out like a rash on the face of the country, covering its populated area at the rate of one and a half major observatory buildings per year from 1836 through the 1850's.

The other special field was that of the biological sciences, which were stimulated by the flora and fauna peculiar to this continent and found in great profusion and in a virgin state through its enormous extent. Thus, from the first explorations onwards there were rich pickings for botanists and zoologists. The special opportunities for ethnology among the Indian peoples and for paleozoology, with the rich stores of the remains of dinosaurs and other fossil animals also created much excitement. All these things gave zest and gusto to American science and its practitioners; the grandest manifestation was the foundation of great museums of natural history, another region in which scientist and artisan found important place.

Lastly and, as in Europe, far from least, rather underlying the whole structure, one finds the clockmakers. It is no coincidence, on the basis of this theory, that the prime exhibit of Yankee ingenuity resulted from the work of the Connecticut clockmakers. Thanks to the fortuitous circumstances that antique clocks are prized by collectors and preserved in museums, we now enjoy the advantage of having a goodly stock of original evidence in this one field. There are now almost enough fine studies for the writing of a monographic history based on this evidence alone. Even some of the documentary sources have been published, among them an outstandingly revealing series of shop records of Daniel Burnap (1759–1838), of East Windsor, Connecticut, one of the earliest and most important of the Connecticut clockmakers.[10] This shows, as one might expect, that he was not a plain maker of clocks but, as was absolutely typical of the practitioners, performed all sorts of metalworking and engaged in the supply and repair of compasses and surveying instruments as well.

If I make too much of the clockmakers, it is only because a far less happy state of knowledge exists elsewhere in the detail of the practitioner movement. Any lyrical enthusiasm displayed for the clockmakers must be tempered by dismal despair when one regards the surveyors and makers of scientific instruments. The original objects, instruments which constitute the most valuable form of documentary evidence, lie junked in attics or at best are conserved in a rusted and rotting condition. Only recently have the most conscientious museums realized that here is something vital to American history and taken pains to restore these objects for proper exhibition as something more than mere personal memorabilia to the little men of science, even when their names are becoming better known.

For many of the important specimens this change has been too late. In modern physics, for example, two of the most important American advances stemmed from the famous Michelson-Morley interferometer experiment, which paved the way for relativity, and the Millikan oil-drop experiment, which was fundamental in studying the electron. In both cases, pieces of the apparatus were apparently cannibalized and finally sold as junk metal. Those instruments were famous in the history of science itself, as

10. Penrose R. Hoopes, *Shop Records of Daniel Burnap, Clockmaker* (Connecticut Historical Society, 1958).

well as vital documents in instrumental ingenuity. Earlier pieces, apparently of less scientific value but just as great in the story of the American practitioners, have disappeared so completely that whole classes lie vacant and without a known member. To any lover of antiques who knows not what to collect, I would suggest the acquisition and analysis of as much scientific junk as can be found.

Returning, then, to the general problem of a unifying concept in the history of the great scientific and technological expansion in America, I feel it can be found in an analogue to the practitioner movement which took place in Europe something like a century before. Seen in that light, Yankee ingenuity is only the phenomenon of the ingenious mechanicians in a more advanced scientific matrix. It is, however, the strongest and most active such movement the world has ever seen, and Americans may be justly proud of it.

Precisely the same process had taken place again and again, perhaps first in the Hellenistic world when the earliest complex machines depicted the divine universe by the motions of gear wheels, and simpler machines were used for surveying. It all happened again in the Empire of Islam and in our own medieval period, when new-found knowledge restored the excitement and led to another few batches of ingenious devices.

Lastly, although the science exhibited considerable continuity, the enthusiasm for high technology did not. The practitioner movement came again with a bang in the ultimate renaissance of the European Scientific Revolution, and perhaps you will agree that it was just another flare-up of good old-fashioned Hellenistic Yankee ingenuity that set America on the path that has led to its present state.

CHAPTER 6

The Difference Between Science and Technology

In 1868 young Edison had his twenty-first birthday. During the preceding year he had mustered up those huge ambitions of a poor and uneducated youth and determined to go off to earn his fortune in the fabulous lands of Latin America. He learned Spanish and got as far as New Orleans where a friend managed to dissuade him from his dream. His two companions went on, however, on the boat to Vera Cruz and quickly died of the raging yellow fever. Edison went back to Boston and started reading the works of Michael Faraday. They excited him strangely and very quickly he was taking out his first patent (of 1,097) for an improvement of the electric telegraph.

By a hair's breadth we might have missed the life of this man who became the American dream, the country's most useful citizen, the benefactor of mankind, the prototype of the great inventor. One little book like "A Boy's Life Of Thomas Alva Edison" did more to inspire a whole generation of scientists and engineers than all the science teaching of the schools. It also produced a fantastic number of better

mousetraps that were going to (but never did) make their
inventors rich and famous like their folk-hero. The flood-
gates of invention were opened here just at that time, and
in the century since, we have seen American cars and
tractors, nuclear bombs and power, chemicals and computers, television and telstar, rockets and lasers, and all the
other symbols of sophisticated might come into being. This
has been so much America's century that the rest of the
world is now deeply conscious of a technological gap. Not
only the prosperity and the military strength of nations, but
also their very survival in the modern world, now depend
on their prowess in science and technology rather than in
their holdings in natural resources or reserves of sheer,
crude manpower.

This has now gone so far that a sort of cargo cult of technology has developed. Every underdeveloped nation, though
full of poverty and illiteracy, needs to have a little nuclear
reactor to bring in the magic of the new age. Worse than
this, nations large and small, rich and poor, beset by the
power of planning and expert advice in an age of science,
find that they should be able to prevent wasteful spending
of their precious funds on useless sciences and weak technologies, and instead spend only on those sensible technologies that are just right for them. The important
question has arisen as to how much such technologies can
be imported and how much they need to be home grown.

In the largest and most scientific nations of the world
there is even greater trouble; the burgeoning explosion of
science into our society has gone at such a pace that in the
U.S. and USSR it is quite clear that the nation is running
out of sufficient allocable money and people to keep science
growing in the style to which it has become accustomed.
One ought now to be worrying about the *over*-developed
countries! Within the last fifteen years the U.S. has slipped
from a place of about one-third of the world's science and

has now become about one-fourth; it is not yet a very worrying slip, but the process accelerates and one must get used to the idea that more and more important ideas are going to be developed first by competing countries, and one must look forward to a brain drain from the U.S. (and the USSR too, if they let them out) comparable to that from Britain, or probably even bigger and faster.

Already one can see signs of strain as the lush federal funding for science begins to dry up and serious competition for funds and for people comes from elsewhere. There begin to be clear signs of an awakening antiscientism; it is a revulsion and disenchantment that takes many forms. First comes a vague feeling that science, which used to stand only for good, has now become associated with harm and evil— one thinks of nuclear weapons, of napalm and mace gas, of electronic bugs, biological warfare, and general pollution of our environment.

Then, too, one sees that science begins to sever itself from the general intellectual life. Young Edison could thrill himself to the core by reading, with little preparation, the researches of Faraday on experimental electricity that had been published not long before. At just that time, though, Maxwell was also publishing, and from *that* line science has changed so much that the comparable research front in high-energy physics now involves something called "current algebra," which is known and can be read only by a very tiny and specialized elite of a few mathematical physicists. However clever you are, you cannot pick up current algebra and read it; far too many books are required to prepare the way; you must get a Ph.D. and put in four or five years as a postdoctoral student first. Technology developed from low technology that *could* have been done four thousand years ago, into a high technology.

In fact, science has become tragically difficult all along the line. Today we have fewer teachers with less competence, relatively speaking, than in Edison's day. In 1890 nearly a quarter of the high school students took physics, now only about five percent have instruction in that subject. Conditions are different, the teachers are obviously much better than in olden days, but physics (in any meaningful way) has become much more difficult at an even greater rate.

I hope I have painted a black and uncomfortable picture. I have done it intentionally in the belief that each of you is in some way already sold on science and technology and that you will therefore have moved automatically into a position to defend science and to manage our affairs so as to cure or minimize the evils I have mentioned. In a way, I have been leading you astray deliberately because many of the bad things I have said are due to a very simple, but tragically naïve, confusion that is widespread. I feel it is terribly dangerous to be naïve in this area, and I want to do what I can to cure the naïveté, even if I cannot solve the problems. The confusion I refer to is that between science on the one hand and technology on the other. So easily can we fool ourselves into believing that we know what these terms mean, and almost as easily can we find it obvious that they have a simple relation to each other. Pure or basic science, it is supposed, is the job of understanding nature, and what one then has to do is to apply this science—to make technology which you can then develop as you wish, to bend nature to the will of man (and in a capitalist country at any rate, to make a tidy profit too). Because of this simple model it seems clear that from science flows all these benefits we wish, and the trick is simply that of finding ingenious ways to apply all this knowledge that we have, pushing the knowledge front before us as we go.

This is just the sort of thing that Edison believed. His job was invention, not discovery; that in a way is typical of the sort of difference I want you to think about. Edison was proud of the fact that he could hire chemists and mathematicians if he needed them. They could not hire him. Since then the position has turned round completely. It has become part of the status race and pecking order for physicists and mathematicians to feel superior to chemists, and they in turn to be superior to engineers. Edison can be despised even as a mere inventor and considered not a scientist at all except for his incidental discovery of the Edison effect, which made the vacuum tube possible and so was technology in any case. What is so wrong in any case with being a technologist? Why do we automatically speak of "pure" science as if technology were dirty? It is a calculated put-down, just as "free" world implies that those you are talking to are slaves.

Let us, then, turn from invective and our innate beliefs and hopes to some objective study of what is involved in sorting science from technology—in comparing these two entities, in contrasting them, and in determining their all-important relationship to each other. (I shall avoid altogether the term "applied science" which begs the issue and only introduces additional ambiguity.) We must, I think, agree that in both science and technology we should concentrate our attention upon "research," on the cutting edge of creation where new things are happening. If we know how that works, it is relatively easy to understand how things are among those who labor behind the research front. It is really not quite so easy as I would like to suppose, for most of the people involved work behind the research fronts rather than at them. What is going on, it happens, is that science and technology are the most competitive activities man is capable of; they are much more of a rat race than business and money, for example. The competition to get to the top is acute and the operation is very wasteful, so that large numbers fail. In a strong sense teachers of science are failures who could not get to the research front, and the technicians are inventors who never made it. Fortunately, nature works this to advantage by

supplying a sufficiency of fine people who are more motivated to be good teachers than failed scientists, etc., but
basically there is a differential of status—and, of course, of
salary—that we cannot ignore. The reason this system
works reasonably well is that there are other rewards than
money in the complex that produces scientists and technologists, teachers and technicians. There is a special satisfaction, and this is, of course, one of the keys in motivation
toward a scientific career.

It should be realized, of course, that in science and technology there appear not to be any absolute standards of
creative achievement. A problem that is difficult is so because very few people can come near to solving it; if all
people become more clever or get better computers, the
problem may be solved or become trivially easy. If almost
anybody can do it, it is not really worth doing. Roughly
speaking, in this area we use the word *excellence* to mean
something that occurs once in a thousand people, and
genius as that which comes once in a million.

Science and technology are both highly creative occupations. They both set a premium on those who can combine
thoughts in interesting ways that simply would not occur
to other people. Edison and Einstein can agree completely
that the biggest part of their motivation is indeed "getting
there first, before the other fellows." Contrary to popularly
held beliefs that they are beset by natural curiosity or by
the hope of doing good, it appears from much modern research that it is competition which holds first place as
incentive.

Right at this point is one of the most important and interesting contrasts between science and technology. In
science you know you have beaten the other man to it if
you publish first. By publishing you stake your claim to
private intellectual property. The more openly you publish, paradoxically enough, the more secure your claim that

the property is exclusively yours. In technology it is otherwise. When you make your invention you must patent it, you must protect it from industrial espionage, you must see that it is manufactured and sold long before it can be copied or replaced by some competitor. In technology you secure private property in the usual jealous way—strangely enough, this is so even in socialist countries where the inventions are national and not private property. Look at Russia's treatment of her rockets, for example.

The difference, I think, emerges because even if science is philosophically a process of generalization and invention of laws, nature appears very strongly to act as if there were only one world to discover. What is more, it acts as if it has to be discovered in a sort of striptease fashion, layer by layer. What I mean is this. If Boyle had not discovered his law, then somebody else would have had to do it. As a matter of fact Marriotte did. If Planck had not found his constant, then we should all have been talking about Joe Blogg's constant. Not only does one feel strongly that each fact and theory is lying there waiting to be discovered, but each one when it comes seems to be discovered by several different people racing against each other to get there. This is creative thought of a very special sort. Boyle had problems getting credit for his law totally different from those Beethoven had getting credit for a symphony or Picasso for a painting. Sometimes one finds this very same competition for the same prize in technology, but for the most part there is much more latitude than in science. I have a strong feeling that if the little green men land from their flying saucer and start talking with us, we shall find immediately that their science is very similar to our science. They might know more, they might know something different, but on the whole their Planck's constant must be the same as ours and their world must have acids and bases in the liquid phase too. Science is completely supranational. It must be much the same for the United States and the Soviet Union, for Catholic and atheist, for Planet Earth or the leaders of the furthest galaxy! On the other hand, there is no reason whatsoever why they should have seen an invention of the incandescent lamp. They might have gone on to fluorescent tubes or to fireflies. They might not have motorcars, just as we do not have their sort of flying saucers. Technology is a sort of arbitrary property of a civilization, whereas science, if you have it at all, has to follow what seems to be more a dictate of nature than a property of our brain. At all events, Boyle and Einstein have priority and property problems in their creativity that are not shared with Beethoven and Picasso, or even with the great historians and philosophers.

Boyle and Einstein are forced to this open publication for an eternal archive that seems to characterize science; it is certainly the thing that makes science not only impersonal and objective, but very attractively impersonal for those bright children who are not very good at getting along with other people. The traditional scientist could win his way to fame and respect by this impersonal publication. The lonely child who curled up with a book could beat the other fellows without even seeing them or being seen by them. He could also know Mother Nature and pry her secrets from her. By the way, note that I had to speak of the traditional scientist; he is very different from the scientist of this generation. If the young are not to trust anybody over thirty, the same is true even more unhappily and forcibly for scientists over thirty. Motivations and personalities, the very nature of dedication, have changed completely, and for the better!

For about twenty years now our society has pleaded with the young to be scientists if they possibly can and has given them scholarships and fellowships and grants. In the old days one was dared to be a scientist if one absolutely had to be, for the good of one's own soul. If you had to, you did physics and starved in a garret just like the artists in bo-hemian Paris. What has happened is that society has made science fairly safe for relatively normal people. The older scientists, so to speak, were nuts; they were very highly motivated and they were paid with prestige and acclaim in their own ranks instead of with mere money, with immortal fame among their own elite. Now it has changed. When I first came to this country about twenty years ago, the comic-strip character of superman was a sort of all-American foot-ball player. Within a few years he changed into a sort of all-American nuclear physicist with rays and such things, and I, for one, knew that the ground rules had been a little changed.

So far I have only spoken of the different outputs of science and technology; one might almost use them for definitions of the modes of research. If, when a man labors, the main outcome of his research is knowledge, something that has to be published openly for a claim to be made, then he has done science. If, on the other hand, the product of his labor is primarily a thing, a chemical, a process, some-thing to be bought and sold, then he has done technology. Now let us look also at the inputs as well as the outputs. The input to a scientist must be all the other papers that are produced by his colleagues and their predecessors. It is quite obvious, in fact, if you look at a scientific paper that it is full of footnotes which are citations back to other people's papers—also to textbooks and to papers not yet published—but on the whole it is to previous papers. When one analyzes the citation patterns, one sees that there is a very close-knit structure here. Scientific papers are as-sembled by a process rather like knitting or the way in which pieces of a jigsaw puzzle are held together by inter-locking with their neighbors. Each scientific paper seems to build onto about a dozen previous papers. Another way of looking at it is to say that, roughly speaking, it is like a human family, except that instead of it taking two parents to make a child it takes about a dozen assorted parents— and they move around like a very free society, enjoying such a deliciously complicated setup as a dozen for a quorum, with each combination producing about a child a year.

It is not only science that works like this, but all scholarship. Research in history and in philology and philosophy also works like a jigsaw puzzle. The difference between science and the rest is simply that science grows at such an enormously rapid rate that most of it at any given time has only just been published. It is like most of all the scientists who have ever lived being alive now, or nearly all scientists being very young. This has always been true and it is not true for other brands of scholarship. Science has a trick of being eternally very young and new. Half of everything we know has been found out in the last decade or so, and this has been true for centuries and will be true for at least several decades yet to come. Because of it science grows, so to speak, from a very thin skin of its research front, whereas philosophy and history grow from knowledge that may be quite old; philosophers can still usefully discuss questions that were very well discussed by Plato and Aristotle—they get places and philosophy moves, but it is not such positive and ever-new knowledge as one can obtain in science.

It used to be that scientists learned about what their colleagues did by reading the journals. Actually they used to read books, then things moved so fast they read only papers, then even faster so they read only letters to the editor in their rapid publication journals. Now they are moving so fast that they do not read but telephone each other, and meet at society meetings and conferences, preferably in beautiful hotels in elegant towns around the world. They get by in what are now called "invisible colleges" of little groups of peers. They are small societies of everybody who is anybody in each little particular specialty. These groups are very efficient for their purpose and, somewhere along the line, people eventually write up their work so that graduate students can read it and get to the research front. By the time it gets published, however, it is so old that all the good research juice has been squeezed out of it, so it is not worth reading if you are really in the business at the research front.

Technologists are quite different in their habits. We have already made a point of the fact that at the research front where useful products are being made the very thing they do not want to do is to publish. On the contrary, they want to keep quiet until the later publicity stage when advertising is in order. In fact, the best reading in technology is, of course, the advertisements. It is odd, though, that technologists do want very much to read. Just as Edison needed to have chemists and mathematicians on hand and read his way through whole encyclopedias and even libraries at random, so today the technologist wants to read everything that is going on in case it might be useful to him in making something new and good. One might indeed say that the scientist wants to write but not read, and the technologist wants to read but not write.

I believe that the so-called information crisis is due to this contrast in positions. It would not be a bad situation if the stuff the technologists wanted to read was exactly that which the scientists are writing. It used to be so in Edison's time—he could read Faraday, but he could not have read James Clerk Maxwell, who mathematized Faraday's electrical theory. What technologists want is something very different, in part what they want is only a sort of boiled-down science such as you learn in college in the process of becoming either a scientist or an engineer or technologist. In part, though, what technologists want is something different. Let me read you some illustrative examples from

Edison himself. They date from the period when the intelligence test score was being invented and Edison set up a test. He called it an *ignoramometer,* which should be answered at a level of, say, ninety percent right for anybody who was going to be a good inventor in his workshop. Here are the sorts of things he wanted people to know:

1. How is leather tanned? 2. Where does the finest cotton come from? 3. Who invented logarithms? 4. Where is Korea. [too easy now, say Sikkim]? 5. What voltage electricity is used on streetcars [subways]? 6. Who composed *Il Trovatore* [who wrote *Mary Poppins*]? 7. What weight (roughly) of air is there in a room 30 feet by 20 feet by 10 feet? 8. [not Edison] What is the heaviest non-metal? 9. [not Edison] What is the breaking strength of the human ankle?

The idea of having all this miscellaneous and mostly useless information on hand and not just where you can look it up eventually is that if you know oddities like these and more, then you can make unlikely combinations in a flash and get places the other fellow cannot get. Technologists want science that has been packed down by education and they want all sorts of unlikely things. That, in a nutshell, is why you have to learn good science, and a lot of it, even if you wish to be an engineer instead of a scientist. It is also worth noting that, according to this model, the most useful person, in science as well as in technology, will be the man who can put together unlikely techniques and bits of knowledge. In designing a college career, or even a high school curriculum, it is precisely the bright scientific or technical kid who should be encouraged to spread his knowledge around. If you want to be a chemist, pure or applied, you should also, as Edison did, spread out into things like computers, Chinese, Buddhist literature, mushroom culture, and the geology of Sikkim. Chances are you will be not merely the only person on your block, but the only one in the world with such a combination, and you may spot the clue everybody else has looked for in vain.

Having now defined something about the terrible twins science and technology, we can begin to analyze their relation. Science is a sort of growing jigsaw puzzle with a dozen sexes, and wherever there is a family of knowledge—an annual supply of knowledge—children are produced. Old knowledge gives rise to new at an exponential rate. From time to time new subdivisions of knowledge appear, but the general process goes on without let or hindrance, without fail even in times of poverty and war, without hurrying in times of need. There is, strangely enough, very little man can do to make knowledge come more or less quickly or to make it come in the directions we may wish. The fruit of the knowledge tree has a habit of wanting to ripen in its own good time. I probably exaggerate for dramatic effect, but something like this seems to be going on. Somehow or other, though we wish it very much and have done for years, we are not yet at the stage of knowing enough to make a cure for cancer.

Technology, the other twin, grows, I believe, in a very similar fashion. It is evident to any historian of technology that almost all innovations are produced from previous innovations rather than from an injection of any new scientific knowledge. There is a sort of state of the art in technology which works very much like the research front in science. We do not see it so well just because the technologists are keeping quite rather than shouting from the rooftops as the scientists do. Indeed, I have often felt that one of the prime difficulties in writing the history of technology is that the major job is the antiquarian one of transferring the state of the art at any time into written form. The research front in science already exists in the form of written ideas, so the job of the historian is much easier and less antiquarian.

We have the position, then, that in normal growth, science begets more science and technology begets more technology. The pyramidlike exponential growths parallel each other, and there exists what the modern physicist would call a weak interaction—at the educational level and the popular book and the *Scientific American* stage—that serves just to keep the two largely independent growths in phase. For the most part, technologists use the science they learned at school and from popular acquaintance, and the scientists use the technology that they have grown up with. Only rarely, but then dramatically and making a historical mountain peak, do the twins show a strong interaction. In the seventeenth-century scientific revolution there was a strong flow from the state of the arts of the artisans into the new scientific apparatus, which exploded ancient science and brought in the modern experimental tradition, with its telescopes and microscopes, barometers and thermometers, airpumps and electrostatic machines. In our own generation the industrial revolution has moved to a new level, mainly through physics—and Edison's electricity in particular—where science is finding its way back into technology. For the most part, science has not helped technology much, but now and again you get anomalous and traumatic events like transistors and penicillin. Again one must be careful; these are the grand exceptions, not the rule. Mountain peaks are not typical. You cannot judge all scientists by the standards of Newton and Einstein. You cannot judge the technological impact of science by the case of transistors.

There is no intellectual difficulty in allowing for the most part that science and technology are only loosely connected systems with very different types of people involved for very different motivations aid purposes, and even trainings. There is, however, a moral difficulty that is particularly interesting and important in an epoch where the exponential growth of the overdeveloped countries has begun to reach saturation and maturity. The money is giving out and the nation is beginning to exercise a new caution about what it spends its money on. The usual temptation of scientists at this point is to lie in a most flagrant and bold fashion. There is, indeed, a long and honorable tradition about lying for the sake of pure science. When Archimedes wanted to pursue his pure geometry he asked his uncle, who was the local National Science Foundation, for financial support on the ground that he would be a useful man to have around in time of war. When war came, being a very bright individual like the late Robert Oppenheimer, he started something quite new, unrelated to his pure science, and burned up the enemy fleet. Leonardo da Vinci had the same technique: promise them technology, make good if you must, but really give them the pure learning that you want and you know they will need in the end.

Although one cannot give any strong proof that science is directly applied at any time to make technology, you must, I think, accept it as a matter of trust that without a live tradition of science you cannot engage in technological growth. Do we really have to stoop so low as to lie about it again and maintain that the latest, biggest accelerator will help us make useful things? Do we need to support mathematics for the direct utility? No, not at all. We can adopt a science-for-science's-sake policy, provided we are clear that this can always be justified by the weak but vital link with technology. We need science so that technologists may grow up immersed in it. I do not avoid the intellectual argument that we also do it because it is the most difficult and elegant thing we can do. Like Everest it is there. The question of justification only becomes important because we ask that society pay for it, and there must therefore be some sort of social contract. Some reason must exist for society to pay; in our age, if you spend on that you must go without something else. The tradition of *libertas philosophandi,* the freedom to follow learning wherever it may lead, is now questioned yet again in the way in which it was questioned by the ancient Romans, by the French revolutionaries, and most recently by communist Hungary. They all thought they could junk useless sciences and pay only for the useful ones. Their civilizations and states were visibly ruined by this tragic policy. It cannot be played like that. The reason is the educational process.

An interesting way in which science differs from non-science in the colleges is in the feedback to the education machine. In the non-scientific departments, like history and English, nearly all the people who survive to take a Ph.D., go back into college teaching. In the sciences only about twenty percent are recycled in this fashion, and the other eighty percent are hired by society to do various jobs in research-front science and technology. In the non-sciences what is society paying for? Are they making teachers to train people to be teachers to train people, etc.? No. The end product that is paid for is the particularly large load of teaching to students below the Ph.D. level. Society is paying for the education of its young citizens in culture, and the higher education only exists as a means of reproduction for the teachers. In science it is different. We are not being paid on the whole to reproduce ourselves, which we do (as elsewhere) for love. Our job is clearly to produce the eighty percent. That is why scientists who succeed at their job do not, in general, want to teach the young. They have a quite different stake in society. For every man in the colleges and universities who does research and replicates himself at a rate of exponential growth with fresh Ph.D. students, there are four or so who work in industry or in government, making the things that society wants to buy.

The outcome of this analysis may now at last be perceived in a rational fashion. Each society has to have science, willy-nilly, whether it likes it or not, because that is what our civilization is all about. And the sciences are their own masters, producing new knowledge in proportion to the amounts that we already know. In fact, if we take the basic sciences—physics, chemistry, mathematics, astronomy, biological science—one can find that in every country in the world that has real education at all, and in every state of the union, each of these sciences is being pursued at almost identical levels. In fact, each entity spends about 0.7 percent of its gross national product (GNP) or wealth on scientists. For every one hundred million dollars of GNP or personal income in any country or state, there happens to be about one physics paper, ten chemistry papers per year, and so on. Nearly all countries play the game or do not do it at all.

With technologies it is different. For the highly developed countries it turns out you can spend only up to about four times as much on creating new products. As we have said, for every scientist, the system produces four technologists. The difference is that all countries and states do not have the same mix. States and countries with a lot of mineral wealth, like Texas, put a lot of their eighty percent manpower production into the earth sciences, and they steal them from every other state and country that has a surplus. So it is elsewhere. In technology you can buy what you want up to a set maximum. In science you have to buy, more or less, what nature will give you, in quantity as well as in quality. In science, even though society pays, there is still some sort of impersonal dedication to nature's rules. In technology there is always something more than the competition. You are supplying something that society wants to buy, and you must be careful that it is something that you want to give your life to make. The young person going into technology has a citizen's responsibility to judge where to put his weight. Much more than that, in an age of pressure, all citizens must be clear that they constitute the society which has the power to buy or not to buy the product of any given technology. Revulsion against such things as napalm is not to be leveled at technologists but at the ordinary political processes whereby society decides it wants to buy such a product.

Finally, I must point out that nowhere is the interaction between science and technology more damnably difficult for society than in the region of medicine. Since the reforms by Abraham Flexner, we have had an excellent truce between the science of medical research and the provision of the technology of healing the sick and maintaining the well. Now, quite suddenly, both the science and the technology have exploded, largely because legislators are often sick old men, and anyhow society is always desperately eager to spend more on medicine than anything else. Molecular biology has produced underpinnings for the science of medicine, and suddenly the medical schools are full of researchers scurrying wherever the glorious new knowledge is taking them. At the same time the very affluence of society, its skill in planning, and the efficiency of medicine itself means that we need very large numbers of the medical technologists and their attendant technicians, M.D.'s, and nurses.

I think that what is happening bears close analogy to the recent divorce between physics and engineering, and the gradual loss of status and salary of the engineers. Unfortunately, however, we do not clearly understand the mechanics of scientific careers and education, and we are hesitant to manipulate the technologies with all the political brutality that seems to be needed. It is a classical situation, where we need a technology of administering technology and we do not even have a decent scientific knowledge of the way that science works. I can only suggest that the most urgent need in science teaching and in planning is more intense thought and analysis, not about the facts and theories of science or the technicalities of technology, but about the place of science and technology in science, the history of these things, and also about such naïve and obviously simple things as the relation between science and technology and the difference between them.

CHAPTER 7

Mutations of Science

THE BLACKEST DEFECT in the history of science, the cause of dullest despair for the historian, lies in the virtual absence of any general historical sense of the way science has been working for the last hundred years. For the scientist it is this more than anything else that makes him feel that this subject is an irrelevant sham and at best makes him undertake to produce a chronicle rather than a history, a mere sequence of who did what and when and how.

For the historian, also, this is a most unpopular field. We are all, it seems, prisoners of the petty compartmentalizations of knowledge that blight our educational arrangements. The system dictates that to get any sound historical training you must resign yourself to a neglect, and hence probably a disdain, of things scientific and, of course, vice versa. Happily, a few escape the dichotomy, and we have a small but increasing number of twinned perverts swelling the ranks of historians of science. What these perverts do is naturally dictated by their several professional competences. Are you classical, medieval, or modern? Physicist, chemist, or biologist? We have just the field for you.

Alas, however, if you are a "modern" historian, it is quite evident that the most exciting periods are in the region of the seventeenth-century Scientific Revolution or, if you prefer French history, in the Enlightenment and the Age of Diderot. The only reasonable alternative to this would be to come closer to the present and risk thereby becoming lost in the hazardous jungle of scientific complexity which flourishes in the period after 1850.

Thus, the history of science looks rather like a zip-fastener that cannot be pulled up the last inch. Every tug from the side of history or from the side of science endangers the entire fabric and keeps us in mortal terror that the whole thing will come completely unstuck and lead to a state of affairs other than intended. To close the last inch—the last century of science—we must do more than catalogue discoveries in each science, more than construct a chronicle of each thread in the webbed tissue of independent disciplines of physics, chemistry, and biology. For the purpose of constructing a general history of recent science we must essay one or the other of two superlatively difficult techniques. Either we must pick some aspect of the scene that is suprascientific, rising above the petty detailed happenings in each little pocket of science, or we must pick some tiny vital spot for a microscopic examination that will reveal more of the character of the instant than just its own most limited local manifestations.

The first alternative is still somewhat imperfect; the second is perhaps a better-beaten path, and we shall attempt that method here. There have been so many separate studies of the evolution of modern physics, of chemistry, and of evolution itself that perhaps some lines may begin to be evident in the general pattern. The over-all picture is clearly that of an intensification of all the magnitudes of science. In the last century it has become more densely populated in manpower, more specialized, more diversified in its specialties.

The early nineteenth century saw the rise of scientific abstracts, consciously designed to make accessible the journals and published papers that were now so numerous that no man could read or hope to assimilate them completely. It saw also the rise of specialized journals, created to cope with the attainment of near autonomy by each of the separate disciplines. At the same time, there began the proliferation of professional scientific societies, many of them, unlike the earlier catch-all national societies, limited to one field or area. We have indeed become so accustomed these days to the independence of the disciplines that it is perhaps the commonest error to regard the history of science as a seeking back along each of these individual lines. It is most difficult to maintain the historical eye and appreciate the essential unity of natural philosophy before this period, a unity which is something more than a mere gathering of the distinct modern scientific subjects such as astronomy and biochemistry.

Our first inquiry, then, is into the process by which the unity of natural philosophy became so split. It had partly begun during the seventeenth-century Scientific Revolution, but the modern state of affairs emerged clearly only after the vital force of that revolution had been spent. The exact sciences of astronomy, physics, and mechanics had got off to an early start in Hellenistic times, and in the work of Newton they were taken to a new plateau of perfection that remained essentially stable and was elaborated only inwardly for more than a century.

During that century, chemistry, as the next of the sciences, began its climb from an uncertain rationale of technology toward some scientific status. The key to its progress was the evolution of techniques competent to deal

with the chemistry of gases—a line of research that includes the great names of Lavoisier and Priestley and involves the curiously plausible false theory of phlogiston.[1]
Once the bastion of theoretical chemistry had begun to
fall, it became evident there was a new inner bastion, that
of organic chemistry. Slowly this was brought to yielding
point, a peculiarly new feature of the battle being that it
was fought by a regular unified army and not by individual
skirmishers. There had grown up, most notably in Germany, whole teams of laboratory chemists, related to one
another as master and apprentice and collaborating closely
on a well-knit strategy of attack. It was not, however, an
easy battle: one of its finest protagonists, Friedrich Woeh-
ler, who was instrumental in establishing the crucial link
between organic and inorganic, became so discouraged
shortly afterwards that he swore it was evident that organic
chemistry never could be the fine systematic science
achieved by its twin, and he deserted back to the study of
metals and their compounds.

As organic chemistry was shakily rising, so also were the
biological sciences. To let drop the magic name of Darwin
is sufficient to demonstrate the intensity of the revolution
that he created and that resounded more than any previous
scientific advance in its public repercussions. Although this
was one of the greatest scientific advances ever made, it is
important to realize that it was not a breakthrough but
rather a break-into. At the time when Darwin's theories

---

1. Perhaps the greatest difficulty of the historian of science *qua* historian
lies in acquiring the proper and necessary sympathy for the plausibility
of wrong ideas. The Aristotelian logic of motion, the geocentric planetary
system, and the phlogiston theory, though all now incorrect in the sense
that they have been superseded by bigger and better, more satisfying
theories, were nevertheless all-powerful in their time, full of explanation
and light, giving useful accord with observation and prediction of observation. One of the most sympathetic treatments of the phlogiston theory is
J. H. White's *The History of the Phlogiston Theory* (London, 1932).

were promulgated, the biological sciences comprised barely more than sort of *catalogue raisonné*. The pieces of the jigsaw puzzle were all neatly sorted, and it was Darwin and his contemporaries who laid out the frame and began the job of creating what was virtually a new science. They were hardly breaking down an old pattern or breaking through a wall that hindered vision, except insofar as the extrascientific epiphenomena of their work impinged on established notions of philosophy and theology.

In physics there was relative quiescence in the old order after the death of Newton. The fruit was ripe for picking, and a rich harvest was reaped, without further radical change by the grand advances in mathematical techniques and the gradual mathematicizing of the whole subject. In addition to this there was one important new area—the study of electricity and magnetism.

Look under the two largest piles of dust in the attic of any old physics laboratory and you will find a pair of giant machines of the age just before our own. They are inevitably a very large and massive vacuum pump and an equally enormous generator of static electricity. They are the first and most impressive large engines of philosophical apparatus, and a great deal of the modern history of science can be told in terms of these precursors of the cyclotron and radio-telescope. The pump and its associated apparatus gave the essential familiarity with pneumatic phenomena that proved crucial in the rise of gas chemistry. The static-electrical machine, by its beautiful and impressive effects, directed the attention of scientists into this region and exposed a whole new body of knowledge.

For our present purpose it is sufficient to note that by the last quarter of the nineteenth century, this new electrical science had been brought under control and, thanks to the masterful analysis of James Clerk Maxwell, it had also been mathematicized into respectability as a member of the family of physics. Maxwell's papers orchestrated electromagnetic theory and united optics with the rest. The whole of physics had become rational on the old pattern, and every available type of phenomenon, with only the most trivial exceptions, was fully understood. Physics comprised a complete exposition of all the actions of matter and energy in the nonchemical province. Chemistry had reached a point of being within shouting distance of a complete system for the reactions in which atoms interchanged to form different compound molecules. Biology had become systematic and reasonably complete in its own province, and there was just enough interaction among all these fields for scientists to be satisfied that the world of scientific learning had been split into its reasonable spheres of influence. Thus, by about 1890, all natural phenomena had been divided and ruled and only unimportant problems remained.

It is perhaps the most precious art of the scientist to develop almost a sixth sense, based on deep knowledge of his whole field, that can tell him which researches are likely to be promising and which not. At this time, though most workers could hardly believe their eyes, and the most cautious were full of contrary warnings, it was obviously reasonable to believe that finality was just around the corner. The only hope for future generations would be to measure each constant of nature to an additional decimal place. A quest for accuracy was then much to the fore, chiefly through the demands of the electrical industry for reproducible and precise measures for their wares. As a result of this, the scientists were heavily influenced by the utility of precision rather than the inner excitement of their work.

The transition from the *fin de siècle* state of approaching perfection of science into the turmoil of our present century is, I believe, the most interesting and also the most crucial line to follow if we wish to have an understanding of the process of modern science. If anything can, it is this that may reveal more significance than its purely local record of advances in some special area at some special time.

It all happened, as it turned out, because one of the obviously trivial remaining problems of physics had, concealed in its bosom, a most potent serpent. This field was the study of electrical discharges in gases. Perhaps it is no coincidence that it was just this field that came about as the one type of experiment which could be done with the descendants of the only two varieties of giant machine in earlier physics. If one used the vacuum pump to suck gas out of a vessel, and then employed the electrical machine (or later, the induction coil) to try to make a spark in it, one could achieve the very beautiful and striking result of brilliant-colored lights and curious bands and other phenomena, just right for a series of magnificent demonstration experiments. They are so good that they are still shown to students today.

Now, what is curious in this history is that the main line of that field really was trivial, just as all the best scientists of the day felt in their souls. We know today that the whole matter of electrical discharges in gases is vastly complicated by too many almost uncontrollable variables. We still do not understand it completely in all its aspects, and we cannot predict exactly what will happen to a given gas at some particular pressure when a discharge of some special wave form is passed through it by electrodes of a particular shape located in some special way. In a sense it is still trivial, though only in pure science. In the applied field it has given us all the lurid neon lights and other multicolored free sideshows of any modern city.

But for the experimental physicists of the day, there was nothing much to do but concentrate on such trivia, in the hope that some good might come of this, rather than the crawl toward extra decimal places. One of the most hopeful, perhaps, was J. J. Thomson, of Cambridge. He had been appointed head of the Cavendish Laboratory at the age of twenty-eight, and a few years later, in 1893, published the first authoritative account in English on recent researches into electrical discharges in gases. Over the years he had a series of brilliant failures in getting sense and order into this field. He devised a technique for getting rid of the disturbing effect of the metal electrodes within the tubes; still it was unrevealing. He measured the speed of propagation of discharges; no clue there to repay all the fearful labor of working with and evacuating tubes many yards long.

Yet through it all he was confident that something good must come up. After all, it was only reasonable to suppose that there should be some benefit from the fact that the properties of gases were so much more simple than those of solids. The mathematical theory was available, and if only the gaseous effects of electricity could be rationalized, we should have some basis for a new electrical theory of the constitution of matter. He had picked on this unrewarding field as the most sensitive one through which to make physics go to a deeper level of understanding.

Undoubtedly the most promising line within the study of electrical discharge in gases was the investigation that had been started by Sir William Crookes and that is now very familiar in its modern application—the cathode ray tube that is the central feature of a television set. Crookes had found a series of very interesting properties of the particular set of rays and bands of light which are given off by the negative electrode, the cathode, in an ordinary discharge tube. Because of various fundamental improvements in vacuum pumps, it had been possible to reach lower pressures of gas than hitherto, and these cathode effects became spectacular. Crookes succeeded in showing that the cathode rays (which could be produced in a focused beam by using a hollowed concave plate as cathode) could be deflected by a magnet and could work a little treadmill within the tube. In these actions they seemed to behave like a stream of little particles carrying electricity— perhaps charged atoms of the gas. He suggested that the new and surprising properties of this substance showed that a new "Fourth State of Matter" beyond solids, liquids, and gases had been produced.

This, in general, was the view of most English physicists. It is most indicative of the close teacher-apprentice organization of science at the time that nearly all the German physicists were opposed to it. Almost to a man they supported the contrary view that was proposed by Hertz, the scientist who had produced new radio waves as had been predicted by the theory of Clerk Maxwell. Hertz felt that the cathode rays were probably some further sort of radiation, rather than tangible matter. In fact, his assistant, Lenard, obtained decisive proof when, by the most ingenious experimental device, he was able to bring the cathode rays outside the tube into the open air. He did it by fitting a thin aluminum window to the tube; obviously, ordinary particles could never pass through anything so solid. But the cathode rays did, and for some little distance outside they showed all their old familiar properties.

The mysterious cathode rays were so much in the news by 1893–94 that many physicists began to turn to them, even if their previous work had been in other, dying parts of the subject. Amid the many good ordinary physicists working thus to resolve the odd situation of German and English camps within the world of science, suddenly in 1895 there arrived an astonishing communication. One of their number, a sound but unremarkable fifty-year-old physics professor at the Royal University of Würzburg, hit on something quite by accident. Because of its importance, I give it in his own words, just as recorded by H. J. W. Dam in a newspaper report about six months later, when he interviewed Wilhelm Conrad Roentgen:

"Now, Professor," said I, "will you tell me the history of the discovery?"

"There is no history," he said. "I have been for a long time interested in the problem of the cathode rays from a vacuum tube as studies by Hertz and Lenard. I had followed theirs and other researches with great interest, and determined as soon as I had the time, to make some researches of my own. This time I found at the close of last October. I had been at work for some days when I discovered something new."

"What was the date?"

"The eighth of November."

"And what was the discovery?"

"I was working with a Crookes tube covered by a shield of black cardboard. A piece of barium platinocyanide paper lay on the bench there. I had been passing a current through the tube, and I noticed a peculiar black line across the paper."

"What of that?"

"The effect was one which could only be produced, in ordinary parlance, by the passage of light. No light could come from the tube because the shield which covered it was impervious to any light known, even that of the electric arc."

"And what did you think?"

"I did not think: I investigated. I assumed that the effect must have come from the tube, since its character indicated that it could come from nowhere else. I tested it. In a few minutes there was no doubt about it. Rays were coming from the tube which had a luminescent effect upon the paper. I tried it successfully at greater and greater distances, even at two meters. It seemed at first a new kind of invisible light. It was clearly something new, something unrecorded." [2]

In all this Roentgen exemplified quite classically the great physicist rather than the chance discoverer. Keeping the business to himself for a while, he put the new effect through its paces and showed that the rays could pass not only through the paper shade but through wood, through thin metals, and even through human flesh, and still cause a glowing of the screen. He showed that other fluorescent substances could make screens, and even that the rays could affect a boxed photographic plate and so draw shadow pictures of all that they penetrated—keys in boxes, bones in the hand. What is so remarkable is not that Roentgen made the accidental discovery but that so many of the people working on cathode rays had missed it. Many such researchers had found their photographic plates in the lab unaccountably spoiled. Sir William Crookes himself had even sent a formal complaint to his suppliers, the Ilford Photographic Company. I wonder if he actually got an apology from them!

Having delayed the announcement of his discovery for a month to extract from it all he could and check the facts, Roentgen made a communication to his local Physico-medical Society at Würzburg (he could do no more than hand the paper in, for all was closed for Christmas recess),

2. The extract from Dam's interview with Roentgen is taken from *McClure's Magazine* for April 1896. It has been reprinted in one of the finest short histories of this period by G. E. M. Jauncey, "The Birth and Early Infancy of X-Rays," *American Journal of Physics, 13* (December, 1945). Dam certainly deserves to be remembered as a very efficient and early pioneer of the modern breed of science writer.

following it up at a more widely attended Physical Society meeting in Berlin on January 4th, 1896. At about this time, he sent out offprints of his Würzburg paper on a massive scale and reached everybody who was anybody in the world of physics. It stopped them dead in their tracks, and somehow or other the matter caught the attention and imagination of the newspapers and public all over the world.

In a matter of days, rather than weeks, every laboratory in the world was playing with the new Roentgen rays (or X rays, to give them their modern name) and doing this to the exclusion of all else. Everywhere scientists and laymen were captivated by the idea of being able to photograph bones without taking them out. In one of the speediest applications of a pure scientific discovery on record, the physics laboratories had become crowded within a week with physicians bringing in patients to check their various real and suspected fractures. In all enthusiasm, as many as could be handled were subjected to half-hour or even hour-long exposures (radiation hazards were then not thought of!) to the frightening accompaniment of the buzzing induction coil and the Roentgen tube glowing with its full hundred candlepower.

It is a pity that it has been forgotten that the discovery of X rays became the first modern scientific break to get banner headlines in the newspapers. Its coverage exceeded that of Charles Darwin: perhaps newspapers had become more sensational in the few intervening decades. It almost rivals, too, the sort of sensation created in our own age by the first atom bomb and the manmade satellite. For weeks, running into months, there were stories, some partly true, some fantastic. The public was fascinated, often for the wrong reasons. Old ladies went into their baths fully clothed, being convinced that the scientists now had mystery rays that could look through brick walls and round corners. From this new mythology of science were born all the wonderful tales of death rays and other science-fictional flights of fantasy, vintage Jules Verne.

In the more serious world of physics, there was equal turmoil. Chance followed chance. In Paris a young man, Antoine-Henri Becquerel, scion of a family distinguished scientifically for three generations, had been working with his father on the subject of phosphorescence, preparing for him beautiful crystals of double salts of potassium and uranium that glowed with the most brilliant light. When X rays came, Becquerel was immediately intrigued by the way in which they were associated with the shining phosphorescent glow of the walls of the Roentgen tube and, getting out his old crystals, he tried introducing them into the tubes to increase the phosphorescence, and perhaps thereby to increase the intensity of the X rays and show that they might be understood perhaps as a hitherto unnoticed effect of strong phosphorescence.

It did not give any satisfying results, because the effect of the X rays was too distracting anyway. Becquerel then tried the various crystals alone, putting them over a wrapped photographic plate to see if they would take their own picture. When none of them worked, he decided to take the most powerful one—the potassium-uranium salt—and expose it to strong sunlight and let it remain on top of the photographic plate for many hours. This worked beautifully, and eventually he found, to his surprise, that the sunlight was unnecessary.

Further work soon led him to conclude that it was the uranium that had the quite fantastic property of giving out radiation—like the X rays, but continuously, without any need for external power or anything artificial. This was the first discovery of radioactivity. From here the subject proceeded by leaps and bounds to the justly famous work of Pierre and Marie Curie. Matter, ordinary inert matter, had been found to be giving off fantastic quantities of heat and light and powerful rays, quite in conflict with all reasonable laws of stability and the conservation of energy. Becquerel's discovery, within the year after Roentgen's, was bad enough. Radium was the last straw.

Also within the year, there were new discoveries proceeding from that of X rays by lines that were more direct, due to a straight follow-through rather than the overtime workings of chance. J. J. Thomson's reaction to Roentgen's announcement was typical of the narrow specialist who can see everything only in terms of his own interests: "I had a copy of the apparatus made and set up at the Laboratory, and the first thing I did with it was to see what effect the passage of these rays through a gas would produce on its electrical properties." To his intense delight and surprise, he found it had exactly the effect he wanted most. It made the gas a good conductor of electricity—in modern terms, it ionized it—and allowed him to experiment without breaking down its electrical resistance by the use of sheer force, as when one normally makes a spark or other discharge.

After this Thomson was well away from the starting post and, using the newly won techniques and reverting to the battle of cathode rays, German or English, he was able to produce definitive proof that they were little charged particles, just as his school had always thought. It turned out that these particles could be measured, and within a year of Roentgen's discovery, Thomson was suggesting that the new corpuscles must be smaller than the smallest known atoms, and carriers of an electrical charge in such a way that they might be the ultimate atoms of electricity that had been postulated long ago by Faraday. The corpuscles moved faster than any atoms, and in proportion to their charge they had a much smaller mass. Thomson had, in fact, discovered the electron, though it took much further work by him and by many others, such as Townshend and Millikan, before the new particle was quite securely established.

Thus, within about two years of Roentgen's accidental discovery, the whole world of physics had split open. For any one toiler in the vineyard, awareness of the change must have been much more sudden and traumatic. At the outset, the whole of science was proceeding toward a foreseeable finish in a number of separate and well-established departments of learning. At a date which was later by perhaps only weeks or days, it appeared that new and unknown rays were waiting to be investigated in all their physical and biological effects. Matter was no longer stable outside the normal reactions of chemistry, and the almost holy law of conservation of energy was being flagrantly violated. Atoms were not the final and ultimate smallest building blocks of the universe, but still tinier, tender particles existed, linking the previously distinct realms of matter and of electricity.

Thus, not only the immediate field of physics had suffered mutation, but chemistry and biology as well were noticeably changed by the consequences of events around 1896. Whereas it had seemed before to chemists and biologists as though their own subjects were developing nicely and firmly grounded on the successes of physics, now they perceived that almost anything might happen in physics and perhaps in their own fields too. The whole population of science became suddenly rather carefree and excitable, and in fact, the first and numerous generation of giant physicists of modern times came out of this particular vintage year of science.

In physics it was the time that a young New Zealander, Ernest Rutherford, came to Cambridge as a research student to work at a new means of detecting radio signals.

Fortunately, young Rutherford, who wanted the financial returns from a patent as a means of bringing his fiancée from New Zealand to England, took the advice of Lord Kelvin to change his plans. Kelvin, hero of the Atlantic Telegraph, took a dim view of radio and claimed that it would be useful only for lightships and other stations that could not use cables, so he recommended that the lad change fields—in 1896. Rutherford did, and perforce took up radioactivity and set it straight by elucidating the alpha rays and atomic disintegration. If it had not been for the vested interest of Kelvin, we might well have had television some decades earlier and the atomic bomb some decades later.

It took some seven or eight years before physics stopped frisking like a newborn lamb. The traumatic end to this phase came as a climactic episode that remains unique in the annals of science. The honors of X rays had been well shared nationally: Roentgen in Germany, Becquerel in France, J. J. Thomson in England, all were among the first winners of the Nobel Prize. American science was just then finding its own feet by importing men from all these schools and by sending its own most promising young men to study with the rapidly growing teams of laboratory workers flushed with the new enthusiasm. Only in this way did American science eventually acquire a stature commensurate with its extraordinary bulk and richness of practitioner activity, and thus the path was set for a new level of attainment.

In the new situation of high activity and tight teams of workers, there was a natural increase in personal and national rivalry. Priority claims became the order of the day, and many of the general scientific journals with weekly publication that we have today were started in this period to provide the newly needed facilities of rapid publication and first claims to ideas.

In France, as everywhere else, X rays remained the dominant influence. At the University of Nancy, the professor of physics was René Blondlot, born in 1849 (just four years after Roentgen) of a father who had been a professor of science before him. His career, like that of Roentgen, was solid but undistinguished until the great discovery. In 1903, the same year as the work of Becquerel, he published a paper that had sprung from his previous activity in measuring the speed of X rays and proving, by great experimental ingenuity, that they were true electromagnetic radiation like light and radio waves.

Using the fact that a spark was affected by X rays and made brighter in their presence, Blondlot managed to detect by this means the expected phenomenon of polarization in X rays, analogous to the polarization of light. He discovered that just as crystals would turn the direction of polarization of light, quartz and lump-sugar rotated the plane of X rays and even of the secondary and tertiary rays thought at this time also to be given out by the Crookes tube.

That was on February 2. On March 23, Blondlot struck again; he was on to something good. Using the spark, he succeeded in showing that all X rays were automatically polarized on emission, and that not only could they be rotated, but also they could be refracted by quartz prism to form a spectrum, just like light. Further, they could be focused by a quartz lens. This was almost too good to be true, but better was yet to come. He noticed that the power fed to the Crookes tube could be turned down so low that there was no phosphorescence, and therefore presumably no X rays, and it would still exhibit all these phenomena detectable by the spark.

Within a few weeks he was back again. This time, having suspected that his new rays were more like infrared rays given off by an incandescent gas burner, he tried such a source, with good positive results. The rays from the lamp would pass through paper, wood, and metal, just like X rays, and still affect his spark. In a subsequent paper, he showed that the new rays were not quite infrared either, and at this point he christened them "N rays," after the University of Nancy, where the work was being done.

In paper after paper Blondlot mined the mother lode. All the phenomena showed by his fluctuating spark could be found in N rays given off by hot bodies or by the sun, and even by certain substances that had only been exposed to hot bodies or sunlight. He found that even the spark was not necessary, for exactly the same effect could be had from the apparent changes in brightness of a dimly illuminated sheet of paper or a spot of some dull phosphorescent chemical. N rays could be produced not only from a body that had been illuminated, but also from one that had been strained by compression or hardening, like a steel file.

While all this was going on, other scientists in France were flocking to Blondlot's banner. They were quite a distinguished band of physicists, including the best in the land, among them Jean Becquerel, son of the Henri of radioactivity, and also the more dubious A. Charpentier, who had been involved with experiments on hypnotism that were then reasonably thought to be quite fantastic. After Charpentier had shown that N rays were given off by all living matter, another man came along and claimed priority for the whole affair, since he had proposed years previously that all life emitted an aura of radiation. Blondlot had all the usual troubles of this sort, but this particular claim was passed to the medical section of the Academy and there left to lie on the table.

Once the start had been made the new field grew rapidly. Nearly one hundred papers on N rays were published in the official French journal *Comptes Rendues* during 1904, representing not only the product of Blondlot and his pupils and assistants but also of other teams of workers growing up in Paris and elsewhere in France. Something like 15 per cent of all physical papers in the journal in this period were on this subject. A success so resounding could not go unrewarded, and eventually, in the same year, the French Academy decided to honor their new discoverer with the considerable Leconte Prize of 20,000 francs and a gold medal.

As it happens, Blondlot had troubles greater than meeting a priority dispute for his claims. As with the work of Roentgen, physicists in all countries had been eager to try out the new effect. In this case, alas, it appeared that not a single Englishman, American, or German could detect satisfactory results. At first there was just disbelief of Blondlot, and the effect was attributed to a mere optical illusion due to the great difficulty of being certain of anything so subjective as small changes in brightness of a dim spark or patch of light in a darkened room. But later the effect had been worked on successfully by many French scientists in many laboratories. Further, when the matter was raised, Blondlot was able to meet the argument by succeeding (as he had not at first) in photographing the change in brightness of the sparks. This enabled him to submit quite absolute objective evidence.

With the consequent increase in perplexity, more scientists abroad tried the experiments, some of them spending much time and ingenuity in trying to get an effect. Some few in countries other than France were indeed successful, but for every one of these there were a dozen men of high repute who became convinced that something was very rotten in the state of French physics. At a summer congress in Cambridge, a number of these men were unofficially brought together. One who felt most strongly was Rubens, of Berlin, a pioneer in the study of infrared rays, upon which Blondlot had dared to touch. Rubens was sweating under a command from the Kaiser to come to Potsdam and provide a demonstration to show that German science was not to be outflanked by French.

From the discussion arose the clear consensus that a first-class physicist, adept in the art of detecting frauds, should go to Nancy and pry into the matter. There was only one man in the world who fitted this bill perfectly, and he was duly unofficially elected to volunteer for the job. The adept was Robert W. Wood, Professor of Physics at Johns Hopkins in Baltimore, one of the most ingenious optical experimenters of all times and the exposer of countless frauds of spiritualist mediums and other perpetrations.

When Wood went to Nancy, he was regaled by a comprehensive demonstration lecture by Blondlot himself. All the great experiments were exhibited. Blondlot demonstrated how a hardened steel file held near one's eyes made a dimly lit clock visible enough to tell the time. He showed the culminating experiment of the N-ray spectroscope with its aluminum prism and lenses which spread the rays into a spectrum and allowed their wave lengths and other optical properties to be analyzed. While doing this, the assistants had become suspicious that Wood was no innocent bystander but had interfered with the apparatus. They repeated the experiment, watching Wood carefully, and suddenly putting on the light when Wood had gone up to the apparatus in the dark and then left it. But all was in order, and the visiting American left amicably after the demonstration.

Next morning, in a letter from Wood to the weekly "priority" paper *Nature,* the beans were well and truly spilled. In the first experiment, when Wood had been allowed to hold the steel file near the eyes of Blondlot, he had secreted it behind his back and held instead a wooden ruler over the forehead of the master. Nevertheless, although wood was one of the few non-emitters of N rays, the experiment had worked completely and Blondlot had seen the clock or not seen it in proper sequence. Wood said that when he tried it himself, no effect whatsoever could be seen. Then, in the big spectroscope experiment, sure enough, the wily American had spirited away the aluminum prism at the beginning of the experiment and sat with it in his pocket throughout the entire success. On the second occasion, when he had already surreptitiously replaced the prism, he fooled the assistant by making as if to do it again but in fact doing nothing.

Wood's letter to *Nature*,[3] a masterpiece of tongue-in-cheek restraint, had a devastating effect. In it he showed it reasonable to attribute all the subjective effects to wishful thinking and to the overpowering difficulty of estimating by eye the brightness of faint objects. The evidence of the objective photographs, he also showed, depended completely upon the easily upset skill of the observer in moving the screens and timing the duration of the very variable flickering spark. From that day onward, there were no N rays. A few papers came out after the fateful September 29 on which that issue of *Nature* appeared, but it seems that they had all been submitted to journals with longer lags between submission and publication, and the authors had failed to retrieve their manuscripts and their reputations.

Blondlot appeared in print only once more. It was another three months before the annual meeting of the Academy at which he was presented with his Leconte prize and gold medal. In a speech which was the epitome of diplomacy, the president, Poincaré, reported that the honor was bestowed for the recipient's entire scientific

3. *Nature, 70* (1904), 530.

work rather than for the N rays in particular. He added that it so happened that circumstances had not allowed all members to acquire that conviction in this matter which could be lent only by personal observation. It was perhaps a belated attempt to acquit French science as a whole, for though some individuals had failed to reproduce the experiments, the movement had been too large in France for much opposition before the grand debunking.

So it was that French science suffered a mortal blow. It took all the prestige of Becquerel and the Curies to effect a restoration of morale. Perhaps the hero-worship of Madame Curie herself was in part not only a tribute to her true worth and her value as a specimen of scientific womanhood but also as an analgesic at a period of traumatic shock for a nation so sensitive to honor. Poor Blondlot was not heard of again. He reached the age of retirement from the university faculty soon afterwards and lived out the rest of his life in Nancy, dying in 1930 after years of obscurity and ill health.[4]

Perhaps it is good that scientists keep always before them the banner of past successes and prefer to forget the few

4. While this book was in course of publication there appeared an excellent analysis of the story of Blondlot and his rays; Jean Rostand, *Error and Deception in Science* (New York, 1960), pp. 13–29. The only previous accounts had been that given by Cohen in *General Education in Science*, eds. I. Bernard Cohen and Fletcher G. Watson (Cambridge, Mass., 1952), p. 87, n. 19 and in the biography of Wood by William Seabrook, *Doctor Wood, Modern Wizard of the Laboratory* (New York, 1941), pp. 234 ff. Blondlot's original papers, translated by J. Garcin, were published in book form by Longmans Green, London, in 1905. Since this story is so worthy of preservation as a pathological example of deviation from the single-minded way in which the cold logic of modern science is often thought to achieve its goals, it would be worthwhile to pursue the facts a little further. Blondlot never told his side of the story. He gave his prize money to the town of Nancy for the purchase of a State Park (which still exists) and made no public statements. He might well have thought that Wood's dramatic treatment of him was unfair and not in the best interest of science.

occasions on which science has taken a wrong turn. The historian, on the other hand, may learn more of the true spirit of science from the pathological example of an inglorious failure than from any normal progress. There are, however, not many good failures to talk about. That which springs to mind most readily is the story of phlogiston in early chemistry, but here the nub of the business is different. Phlogiston was plausible and, in a sense, in keeping with available observations of the chemistry of gases. It was a reasonable explanation—until, as more experiments accumulated, the properties of phlogiston had to be stretched so that it became so general and all-pervading and omni-propertied a gas as to be useless as an explanation.

The curious error of N rays is much more a sort of mass hallucination, proceeding from an entirely reasonable beginning. By no means can it be considered as any sort of hoax or crank delusion—it was a genuine error. It mushroomed into a complex that could have been possible only in that short and glorious epoch when physics had suddenly found the first great massive breakthrough in its modern history. Out of that arose the whole science of radioactivity, of atomic physics, and eventually all the material of particle physics.

Oddly enough in an age that has produced the new wonder of atomic bombs and energy, many physicists now feel that in some ways physics is once more near a point at which the end is almost in sight—for the theory, if not for the world. The present maze of fundamental particles is getting to that stage of complexity where it can be resolved only by some master stroke. It seems likely that such a stroke may also be closely linked with the basic problem of establishing some unified theory in which relativity and the quantum theory appear as separate facets or consequences of the same simple thing. This, if effected, would bring much of physics to a complete and desired end—perhaps. On quite different grounds, if quantum theory decrees a fundamental limit of fineness in our observations, and if the size of the universe is limited and not infinite, then it follows in some way that science is also necessarily limited and finite and that completeness of some sort is inevitable. This is meat for the philosophers rather than the historians, and apparently the physicists are not yet worried by that end-of-century, end-of-science feeling they had before the mutation of 1895.

One may say, however, that the first atomic explosion in history was not in 1945; it took place exactly half a century earlier. And in 1895 it was not some mere laboriously built artifact of science that exploded but rather the science itself. Our modern world is largely the result of efforts to piece together the fragments left by that traumatic and crucial explosion.

CHAPTER 8

Diseases of Science

THE USE of a mathematical and logical method is so deeply embedded within the structure of science that one cannot doubt its power to bring order into the world of observation. Perhaps the best classical statement of this is given by Plato in his *Laws*, where he remarks that "arithmetic stirs up him who is by nature sleepy and dull, and makes him quick to learn, retentive and shrewd, and aided by art divine he makes progress quite beyond his natural powers."

This is amply demonstrated by the rich return whenever the scientific methods of measurement and mathematical treatment have been used, be they within the sciences as in biology, or in human affairs as in economics and other segments of what was once called political arithmetic. It does not, of course, follow that quantification followed by mathematical treatment is in itself a desirable and useful thing. The pitfalls are many; for example, it is almost certainly an arbitrary if entertaining procedure to grade the various geniuses that the world has seen and give them so many marks out of a hundred for each of the qualities they have demonstrated or failed to demonstrate.

161

Now the history of science differs remarkably from all other branches of history, being singled out by virtue of its much more orderly array of material and also by the objective criteria which exist for the facts of science but not necessarily for the facts of other history. Thus, we can be reasonably sure what sort of things must have been observed by Boyle or Galileo or Harvey, in a way that we can never be sure of the details of Shakespeare's life and work. Also, we can speak certainly about the interrelations of physics, chemistry, and biology, but not so positively about the interdependence of the histories of Britain, France, and America.

Above all, there is in the field of science a cumulative accretion of contributions that resembles a pile of bricks. Each researcher adds his bricks to the pile in an orderly sequence that is, in theory at least, to remain in perpetuity as an intellectual edifice built by skill and artifice, resting on primitive foundations, and stretching to the upper limits of the growing research front of knowledge.

Now, seemingly, by means of the art divine of arithmetic, an array so orderly is capable of some sort of exact analysis which might progress beyond the natural powers afforded us by the usual historical discussions. It is perhaps especially perverse of the historian of science to remain purely an historian and fail to bring the powers of science to bear upon the problems of its own structure. There should be much scope for a scientific attack on science's own internal problems, yet, curiously enough, any such attack is regarded with much skepticism, and the men of science prefer, for the most part, to talk as unskilled laymen about the general organizational problems with which science is currently beset.

Fortunately, it happens that the most revealing issues in the history of the last few centuries of science have much in common with the basic problems currently afflicting the structure and organization of science. Both considerations concern what one might well call the "size of science"— the magnitude of the effort in terms of numbers of men working, papers written, discoveries made, financial outlay involved. For the history of science, the treatment of such magnitudes by a process of refined head-counting and suitable mathematical manipulation may provide one much-needed way of viewing the forest of modern science without the distraction of the individual trees of various separate technicalities. Provided only that we take the precaution to link the results at every possible stage with such information as we have already gleaned from purely historical considerations of the evidence, it might do much to amplify that evidence. It is in a very similar way that economic history can augment social history and provide a new and more nearly complete understanding of processes that previously were only partly intelligible on qualitative lines.

Before entering this region, I must post a caveat with respect to the claim that such an analysis might have direct bearing on our understanding of present problems and future states of science. Whatever our reasons for accepting the study of history as a legitimate and valuable activity of scholars and teachers, one of the claims not customarily made is that of direct utility. We do not advise that a good grounding in history can make one an efficient politician. We do not maintain that the historian is the possessor of any magic crystal ball through which he can look into the future. If I suggest that the history of science is perhaps more useful than most other histories, it is only because of the peculiar regularity and verifiability of its subject matter. Since such oddities exist, however, it is useful to stretch the method to the full and examine critically any benefits which might thereby accrue.

For a preliminary exercise in the internal political arithmetic of science, let us first examine the history of the vital process that made science assume a strongly cumulative character. The origin of this was in the seventeenth-century invention of the scientific journal and the device of the learned paper—one of the most distinct and fundamental innovations of the Scientific Revolution. The earliest surviving journal is the *Philosophical Transactions of the Royal Society of London,* first published in 1665.[1] It was followed rapidly by some three or four similar journals published by other national academies in Europe. Thereafter, as the need increased, so did the number of journals, reaching a total of about one hundred by the beginning of the nineteenth century, one thousand by the middle, and some ten thousand by 1900. According to the *World List of Scientific Periodicals,* a tome larger than any family Bible, we are now well on the way to the next milestone of a hundred thousand such journals.

Now this provides a set of heads that are reasonably easy to count. For the earlier period there exist several lists giving the dates of foundation of the most important scientific serial publications; for more recent years we have the *World List* and similar estimates. Of course there is some essential difficulty in counting *Physical Review* as a single unit of the same weight as any *Annual Broadsheet of the*

1. The most readable recent account of the genesis of the Royal Society and its *Philosophical Transactions* is Dorothy Stimson, *Scientists and Amateurs* (London, 1949). For the other national societies, the standard secondary source is Martha Ornstein, *The Role of Scientific Societies in the Seventeenth Century,* 3rd ed. (Chicago, 1938). The only good general history of the later history of the scientific periodical is a short article by Douglas McKie in "Natural Philosophy Through the Eighteenth Century and Allied Topics," Commemoration Number to mark the 150th Anniversary of the *Philosophical Magazine* (London, July, 1948), pp. 122–31. See also John L. Thornton and R. I. J. Tully, *Scientific Books, Libraries and Collectors* (London, 1954), especially Ch. 8, "The Growth of Scientific Periodical Literature," which cites several further references.

Society of Leather Tanners of Bucharest, but for a first order of magnitude, there seems no overriding difficulty in selecting which heads to number.

If we make such a count extending in time range from 1665 to the present day, it is immediately obvious that the enormous increase in the population of scientific periodicals has proceeded from unity to the order of a hundred thousand with an extraordinary regularity seldom seen in any man made or natural statistic. It is apparent, to a high order of accuracy, that the number has increased by a factor of ten during every half-century, starting from a state in 1750 when there were about ten scientific journals in the world. From 1665 to 1750, the birth span of the first ten journals, the regularity is not quite so good, but this indeed is exactly what one might expect for a population that was then not large enough to treat statistically. No sort of head-counting can settle down to mathematical regularity until the first dozen or so cases have been recorded.

The detail at the beginning of the curve of growth is rather revealing in terms of its historical implications. Starting in 1665, the curve proceeds for a couple of decades as if there had been healthy growth. By that time, the growth acts as if it had started from a first journal at a date nearer to 1700 than 1665. Thus, the curve indicates that, in some sense, the scientific journal was born a little too soon. The first publications were demonstrably precursors rather than true originators of the process. This is particularly interesting when one considers the difficult periods which the Royal Society and the other academies experienced once the initial flush of enthusiasm had passed. They went through grave crises and had to suffer rebirth early in the eighteenth century.

In the course of this proliferation of the scientific journals, it became evident by about 1830 that the process had reached a point of absurdity: no scientist could read all the

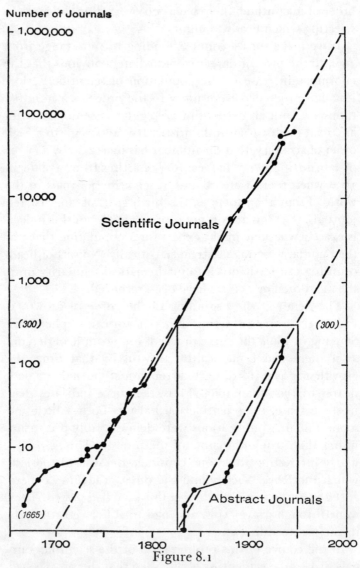

Number of Journals

1,000,000

100,000

Scientific Journals

10,000

1,000

(300)   (300)

100

10

(1665)

1700   1800   1900   2000

Abstract Journals

Figure 8.1

Number of journals founded (*not* surviving) as a function of date. The
two uppermost points are taken from a slightly differently based list.

journals or keep sufficiently conversant with all published work that might be relevant to his interest. This had, in fact, been an attendant worry from the very beginning of the operation, and the first duty of the earliest journals was to review all published books and all papers which had appeared in the organs of the other national academies. But by about 1830 there was clearly trouble in the learned world and, with an assemblage of some three hundred journals being published, some radically new effort was needed. Yet again there was an invention as deliberate and as controversial as the journal itself: the new device of the abstract journal appeared on the scene.

Now a single abstract journal could never suffice, and in accordance with the convenient compartmentalization of science current by this time, further abstract journals were created to fill the needs of the various specialist groups. Because it presented a solution to the crisis, the abstract journal removed the pressure, and the number of plain journals was enabled to grow unhampered. This growth has continued to the present day. On account of this proliferation, however, the number of abstract journals has also increased, following precisely the same law, multiplying by a factor of ten in every half-century. Thus, by about 1950 we reached the point at which the size of the population of abstract journals had attained the critical magnitude of about three hundred. This is, of course, the reason why during the last decade scientists have been concerned about the need for abstracts of abstracts, calling this an "information problem" which seems to require some process of electronic sorting of abstracts as a means of coping with the rising flood of literature.

It is interesting to reflect that, on the basis of this historical evidence, one can show that any new process would bear the same relation to abstracts as the abstracts have to original papers. This relation involves a compression by a factor of about three hundred—the number of journals that seem to have necessitated the coming into being of each abstract journal.

Now it seems that the advantage at present providable by electronic sorting may be of a considerably smaller order of magnitude—perhaps a factor of the order of ten. If this is so, it follows that the new method must be no more than a palliative and not the radical solution that the situation demands. It can only delay the fateful crisis by a few paltry decades.

The seriousness of the crisis is evident from the change in form and function of physics papers in recent years. Collaborative work now exceeds the single-author paper, and the device of prepublication duplicated sheets circulated to the new Invisible Colleges has begun to trespass upon the traditional functions of the printed paper in a published journal.[2] If we do not find some way of abstracting the abstracts, it may well happen that the printed research paper will be doomed, though it will be difficult to rid ourselves of the obsession that it seems vital to science.

2. The new Invisible Colleges, rapidly growing up in all the most hard-pressed sections of the scholarly research front, might well be the subject of an interesting sociological study. Starting originally as a reaction to the communication difficulty brought about by the flood of literature, and flourishing mightily under the teamwork conditions induced by World War II, their whole *raison d'être* was to substitute personal contact for formal communication among those who were really getting on with the job, making serious advances in their fields. In many of these fields, it is now hardly worth while embarking upon serious work unless you happen to be within the group, accepted and invited to the annual and informal conferences, commuting between the two Cambridges, and vacationing in one of the residential conference and work centers that are part of the international chain. The processes of access to and egress from the groups have become difficult to understand, and the apportioning of credit for the work to any one member or his sub-team has already made it more meaningless than before to award such honors as the Nobel Prize. Are these "power groups" dangerously exclusive? Probably not, but in many ways they may turn out to be not wholly pleasant necessities of the scientific life in its new state of saturation.

The most remarkable conclusion obtained from the data just considered is that the number of journals has grown exponentially rather than linearly. Instead of there being just so many new periodicals per year, the number has doubled every so many years. The constant involved is actually about fifteen years for a doubling, corresponding to a power of ten in fifty years and a factor of one thousand in a century and a half. In the three hundred years which separate us from the mid-seventeenth century, this represents a factor of one million.

One can be reasonably surprised that any accurate law holds over such a large factor of increase. Indeed, it is within the common experience that the law of exponential growth is too spectacular to be obeyed for very long. Large factors usually introduce some more-than-quantitative change that alters the process. Thus, if only the Indians had been wise enough to bank at compound interest the small sum for which they sold the island of Manhattan, it would now, at all reasonable rates of interest, have grown to be of the same order of magnitude as the present real estate value of that area.

Now not only is it therefore quite exceptional that anything could have grown so regularly from unit size to the order of hundreds of thousands, but it is altogether remarkable that this particular curve should be a normal, compound interest, exponential law of growth rather than any of the other alternatives that exist, some of them more simple, some more complex. The exponential law is the mathematical consequence of having a quantity that increases so that the bigger it is the faster it grows. The number of journals has behaved just like a colony of rabbits breeding among themselves and reproducing every so often. Why should it be that journals appear to breed more journals at a rate proportional to their population at any one time instead of at any particular constant rate?

It must follow that there is something about scientific discoveries or the papers by which they are published that makes them act in this way. It seems as if each advance generates a series of new advances at a reasonably constant birth rate, so that the number of births is strictly proportional to the size of the population of discoveries at any given time. Looking at the statistics in this light, one might say that the number of journals has been growing so that every year about one journal in twenty, about 5 per cent of the population, had a journal-child—a quotient of fecundity that is surely low enough to be reasonable but which must inevitably multiply the population by ten in each succeeding half-century.

The law of exponential increase found for the number of scientific journals is also obeyed for the actual numbers of scientific papers in those journals. In fact, it seems an even more secure basis to count the heads of whichever papers are listed by one of the great abstract journals or bibliographies than to take a librarian's list of the journals themselves. A list of papers is likely to be a little more comprehensive and more selective than any list of journals which may from time to time publish scientific papers immersed in nonscientific material.[3] As a good specimen of the result

3. In addition to the examples here cited, there are several known to me in standard bibliographies of the sciences and commentaries thereon. For X-ray crystallography there is the graph reproduced by William H. George in *The Scientist in Action* (London, 1938), p. 232, fig. 27 (taken from the bibliography by Wyckoff, *The Structure of Crystals*, 2nd ed. New York, 1931, pp. 397–475). For experimental psychology see Robert S. Woodworth, *Experimental Psychology* (New York, 1938), p. iii. For astronomy there is the monumental work of Houzeau and Lancaster, *Bibliographie Générale de l'Astronomie* (Brussels, 1882), II, p. LXXI. This last is especially remarkable for the fact that it shows the full time-scale, beginning with a slow exponential growth (about forty years for a doubling) and then changing to the modern normal rate (about fifteen years to double) just at the time of the first astronomical periodical publications in the early nineteenth century. For chemical papers see the analysis by Laurence E. Strong and O. Theodor Benfey, "Is Chemical Information Growing Exponen-

of such a statistical investigation of numbers of scientific papers, there is next presented a curve showing the numbers of papers recorded by *Physics Abstracts* since it came into being in 1900. In the earliest decade, this journal's main function was to record electrical engineering papers, and not before World War I did it find it useful to list the physics section separately; we therefore ignore the mixed data before 1918.

It is, however, quite remarkable that from 1918 to the present day the total number of physics papers recorded in the abstracts—clearly a rather complete and significant selection—has followed an exponential growth curve to an order of accuracy which does not fluctuate by more than about 1 per cent of the total. There are now about 180,000 physics papers recorded in these volumes, and the number has steadily doubled at a rate even faster than once every fifteen years. In this curve, one particular side effect is worth noting. The data show that during World War II, in the period 1938–48, the production of physics papers was reduced to reach a minimum of very nearly one-third of what it normally would have been. In the whole decade including the war, some 60,000 instead of 120,000 papers came out.

Two diametrically opposed conjectures have been made with respect to the effect of the war upon science. The one school would argue that the enormous stimulation of giant projects like that of the atomic bomb helped science in a way that no peacetime activities could have afforded. The

tially?" in *Journal of Chemical Education*, 37 (1960), p. 29. Since my first publication on this subject (in *Archives Internationales d'Histoire des Sciences*, No. 14, 1951, pp. 85–93) extended and republished in a more popular form in *Discovery* (June, 1956), pp. 240–3, there have come to my notice about thirty such analyses, all with similar results. It seems beyond reasonable doubt that the literature in any normal, growing field of science increases exponentially, with a doubling in an interval ranging from about ten to about fifteen years.

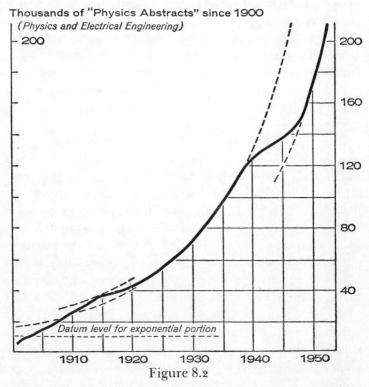

Thousands of "Physics Abstracts" since 1900
(Physics and Electrical Engineering)

Figure 8.2

Total number of *Physics Abstracts* published since January 1, 1900. The full curve gives the total, and the broken curve represents the exponential approximation. Parallel curves are drawn to enable the effect of the wars to be illustrated.

other school says that the mobilization of men and money for purpose of war effort rather than for scientific advance was a diversion, an actual retardation instead of an acceleration of science. The graph shows immediately that neither of these things happened—or, rather, if they did, they balanced each other so effectively that no resultant effect is to be found. Once science had recovered from the war, the curve settled down to exactly the same slope and rate of progress that it had before. It had neither a greater nor a less initial slope; it is exactly as if the war loss had not occurred. The present curve runs accurately parallel to its projected prewar course.

Returning to the main investigation, we can note that once again the accuracy of exponential growth is most surprising, especially because of the large factor involved, and also because its regularity is so much greater than one normally finds in the world of statistics. I might add that exactly the same sort of result occurs if one takes the headcount for scientific books or for abstracts of chemical, biological, or mathematical papers.[4] It may also be found in the bibliographies which exist for particular specialties within any of these domains. One may, in fact, with a suitably documented topic, perform such a mathematical analysis and thereby demonstrate very clearly the successive phases: first, precursors; then, a steady state of exponential growth; next, a decline to linear growth, when no new manpower is entering the field; and finally, the collapse of the field, when only a few occasional papers are produced, or an alternative revival, should it suddenly take on a new lease of life, through a redefinition of its content and mode of operation.

4. The figures for book publication and the size of libraries are the subjects of many investigations, several of them instigated by worried librarians charged with the management of their monster. Perhaps the best selection of data is in F. Rider, *The Scholar and the Future of the Research Library* (New York, 1944). Roughly speaking, both the world population of book titles and the sizes of all the great libraries double in about twenty years (estimates usually range from seventeen to twenty-three years). If we allow that in some five hundred years of book production there must have been some twenty-five doubling periods, this will give about $2^{25} = 30,000,000$ books alive today, a figure conforming well with normal estimates. In the *Third Annual Report of the Council on Library Resources* (period ending June 30, 1959), where such data are presented, I find a wistful comment that deserves repetition: "The world's population is laid to rest each generation; the world's books have a way of lingering on." Such is the stuff of cumulative growth, the distinction of scholarship in general, but of science in particular.

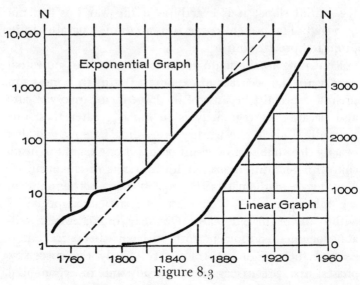

Figure 8.3

Total number of papers published in the field of mathematical theory of determinants and matrices, plotted exponentially (left) and linearly (right). There are three stages in the growth, the first an irregular period of precursors and a slightly premature beginning, from 1740 to about 1800. The next stage is one of pure exponential growth from 1800 to about 1880 and the last is a period of linear growth extending from 1880 to the present. In the exponential portion there is a doubling every twelve years. In the linear portion the growth maintains its value at 1880, i.e., about thirty-five papers per year, or roughly one dozen full-time workers in the field.

So far we only have the very crudest measure of the size of science; there has been no discussion of the relation between the number of papers and the number and quality of the scientists working and the research they produce. It is relatively easy to establish a relationship between scientists and their papers. For example, one can readily take an index volume for several years of publication in a particular journal or over a whole field and count the number of men who published but one paper, those with two, three, and so on. This has been done many times, and for my present purpose, it will suffice to cite Lotka's Law of Productivity, [5] which states that the number of authors publishing just N papers is proportional to $1/N^2$. Thus, if you have a certain chance of producing one paper during your lifetime, you have one-quarter that chance of producing two, one-ninth for three, one-hundredth for ten, and so on.

Again, this is a reasonably expected mathematical law, but it is surprising to see that it seems to be followed to much greater accuracy than one might predict. Once more, it is surprising to find that this seems to be a universal law. Thus, it is obeyed equally well by data taken from the first few volumes of the seventeenth-century *Philosophical Transactions* and by those from a recent volume of *Chemical Abstracts*. The distribution of productivity among scientists has not changed much over the whole three hundred years for which papers have been produced.

As a result of the constancy of this law, it is possible to say that over the years there have been about three papers for every author. If we care to define a scientist as a man who writes at least one scientific paper in his lifetime, then the number of scientists is always approximately one-third of the number of published papers. Actually, the mathematics of this computation is not quite trivial; it is necessary to make a somewhat arbitrary assumption about the maximum number of papers that could be written by any man

5. Lotka's Law was first published in "The Frequency Distribution of Scientific Productivity," *Journal of the Washington Academy of Sciences*, *16* (1926), 317. It is commented upon as an example of an almost universal distribution law in George K. Zipf, *Human Behavior and the Principle of Least Effort* (Cambridge, Mass., 1949), pp. 514–16 (some of the theory and examples are not entirely trustworthy). An independent and more mathematical formulation in terms of skew distribution functions and their theory may be found in Herbert A. Simon, *Models of Man, Social and Rational* (New York, 1957), pp. 160–1, where further source materials are cited.

in one lifetime.[6] Happily, the agreement with statistical data is so good that assumptions do not appear to be very critical.

Having established this, we may transfer all our remarks about the growth of scientific literature into equivalent remarks about the manpower involved. Hence, during the last three hundred years, the size of the labor force of science has grown from the first few to the order of hundreds of thousands. Now this is something so familiar, it seems, from discussion of the explosion of the world population, and from the well-known troubles of libraries, which seem to be doubling in size every few decades that it may look as if we are merely making new soup with old bones.

To state it a little more dramatically, however, we may remark that at any time there co-exist in the scientific population scientists produced over, let us say, the last forty years. Thus, at any one time, about three doubling periods' worth of scientists are alive. Hence, some 80 to 90 per cent of all scientists that have ever been, are alive now. We might miss Newton and Aristotle, but happily most of the contributors are with us still!

It must be recognized that the growth of science is something very much more active, much vaster in its problems, than any other sort of growth happening in the world today. For one thing, it has been going on for a longer time and more steadily than most other things. More important, it is growing much more rapidly than anything else. All other

6. So far as I know, the record for meaningful scientific publications in huge quantities is held by William Thompson, Lord Kelvin. From about 1840 to 1870 he produced about 8.5 papers per year, thereafter until his retirement some 15.0 per year (for a period of about thirty years), then about 5.0 per year until his death in 1907; in all a total of about 660 papers in one lifetime, a working average of one fine paper per month, year in and year out. Almost every one of these could be viewed as a major scientific contribution. See Postscript, p. 195.

things in population, economics, nonscientific culture, are growing so as to double in roughly every human generation of say thirty to fifty years. Science in America is growing so as to double in only ten years—it multiplies by eight in each successive doubling of all nonscientific things in our civilization. If you care to regard it this way, the density of science in our culture is quadrupling during each generation.

Alternatively, one can say that science has been growing so rapidly that all else, by comparison, has been almost stationary. The exponential growth has been effective largely in increasing the involvement of our culture with science, rather than in contributing to any general increase in the size of both culture and science. The past three centuries have brought science from a one-in-a-million activity to a point at which the expenditure of several per cent of all our national productivity and available manpower is entailed by the general fields of science and its closely associated applications.

An excellent example of such concentration is the electrical engineering industry, the technology of which is more implicitly scientific than any other. Published manpower figures show the usual exponential increase, acting as if it started with a single man *ca.* 1750 (the time of Franklin's experiments on lightning) and doubling until there were two hundred thousand people employed in 1925 and an even million by 1955. At this rate, the whole working population should be employed in this one field as early as 1990.

Returning for a moment to the history of the process rather than its statistics, it seems reasonable to identify by name this growth of science and its associated technologies from the small beginning to its present status as the largest block of national employment. It is the process we call the Industrial Revolution, if one thinks in terms of technology, or the Enlightenment, if one stresses the cognitive element.

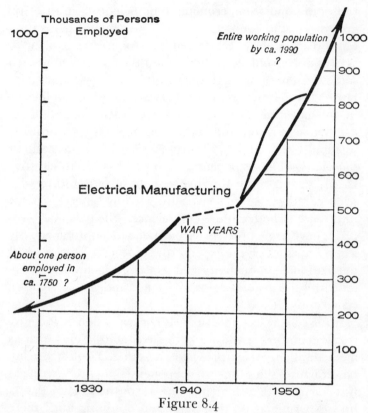

Figure 8.4

Adapted from figures published by *The Manchester Guardian* for March 20. 1956: "The Electrical Industry Today," by Dr. Willis Jackson, F. R. S.

The movement started in Europe in the mid-seventeenth century and reached large proportions measurable by thousands, rather than units, in the late eighteenth and early nineteenth centuries. Thus, our various graphs of cumulative growth may be regarded as charting quantitatively the course of this Industrial Revolution and Enlightenment and providing a key to the various dates and phenomena associated with their progress.

It is instructive in this study to compare the growth charts of Europe with those for the United States. All the available statistics show that the United States has undergone the same sort of accurately exponential proliferation as Europe. The difference is, however, that once the United States started, it made its progress by doubling in scientific size every ten years rather than every fifteen. This was remarked upon already in 1904, in a brilliant essay in *The Education of Henry Adams* (Chapter XXXIV). The explanation of the rate difference is difficult, but the fact seems quite clear. Once the United States had, so to speak, decided to get down to a serious attempt at scientific education, research, and utilization, it was able to carry through this process at a rate of interest considerably higher than that in Europe.

A great part of the explanation is probably due not to any special and peculiar properties of the American way of life as compared with the European but merely to the fact that this country was expanding into a scientific vacuum. Furthermore, it was doing it with the help of that high state of science already reached and held as a common stock of knowledge of mankind at the date when the United States started its process. Europe had to start from the beginning, and by the eighteenth or nineteenth century it had a considerable accretion of tradition and established institutions of science and technology.

Whatever the reason, the United States continued to expand at this rate faster than Europe, and eventually it acquired an intensity of science in society that became greater than that of Europe. One can consider the scientific advancement of Russia in exactly the same way. In Czarist Russia science was not altogether inconsiderable—it partook of the general level of Europe—but after 1918 a determined effort was made to expand science. Again the statistics show that the advance has been very accurately exponential, and that the doubling time is of the order of some seven years rather than the ten of the United States and the fifteen of Europe. Again, one can attribute this in large part not to any particular excellence of the Russians or to a degree of crash-programming but rather to the fact that if they wanted to do the job at all, there was only one way of doing it, and this involved being able to start from a world-state of scientific knowledge that was considerably higher for them than for the start of the United States.

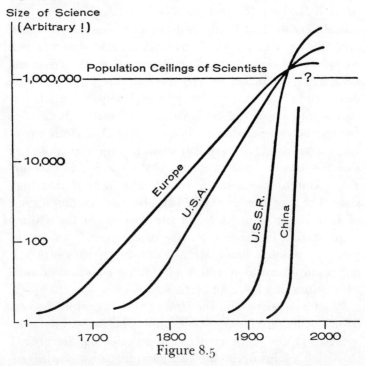

Figure 8.5

Schematic graph of the rise of science in various world regions. The measures, the shapes of the initial portions of the curves, and the way in which the curves turn over to their respective ceilings toward the top are all merely qualitative.

Lastly, we may take the case of China. Here we have an even more recent start and, consistent with the theory, we see that the statistics in that country indicate a doubling every five years. As an indication of this, it has already just become necessary and advisable to prepare running English translations of the chief Chinese scientific journals, as we have now been doing for the Russian literature over some few years. Again, rather than attribute any particular high quality to the Chinese, I would suggest that they are simply expanding into a larger scientific vacuum, starting at a higher level than any of the earlier protagonists.

The whole thing is like a gigantic handicap race in which the country that starts last must necessarily have the highest initial speed, and it seems fairly conclusive that this speed can readily be maintained—it certainly has been by America—so that the state of science must eventually reach the concentration that we see in the most highly developed countries. It is reasonable to suppose from the very universality of science and from its supranational qualities that it is much more likely for the world to reach a state of uniform development and exploitation in this direction than in many another. The handicap race of Industrial Revolutions has indeed been so well designed that it seems likely that all runners will come abreast, reaching a size of science proportional to their total populations, at much the same time, a time not too many decades distant into the future.

Because of the obvious importance of the scientific race between the United States and Russia, and that which may well occur between these countries and China, this study of the natural history of Industrial Revolutions clearly needs more attention. The modern scientific development of Japan would provide an excellent case history. The very slow beginnings in modern India might throw light on what it is that constitutes a true onset of this sort of exponential Industrial Revolution.

Having now discussed the historical origins and statistical progress of the device of the scientific paper and the profession of the scientist, we must next consider the decline and fall of these things. It is indeed apparent that the process to which we have become accustomed during the past few centuries is not a permanent feature of our world. A process of growth so much more vigorous than any population explosion or economic inflation cannot continue indefinitely but must lead to an intrinsically larger catastrophe than either of these patently apparent dangers.

To go beyond the bounds of absurdity, another couple of centuries of "normal" growth of science would give us dozens of scientists per man, woman, child, and dog of the world population. Long before that state was reached we should meet the ultimate educational crisis when nothing might be done to increase the numbers of available trained professionals in science and technology. Again, to take a reasonably safe exaggeration, if every school and college in the United States were turned to the exclusive production of physicists, ignoring all else in science and in the humanities, there would still necessarily be a manpower shortage in physics before the passage of another century.

The normal expansion of science that we have grown up with is such that it demands each year a larger place in our lives, a larger share of our resources. Eventually that demand must reach a state where it cannot be satisfied, a state where the civilization is saturated with science. This may be regarded as an ultimate end of the completed Industrial Revolution. Thus, that process takes us from the first few halting paces up to the maximum of effort. The only question that must be answered lies in the definition of that saturated state and the estimation of its arrival date.

Fortunately, the mathematical theory is again most helpful if we demand only an approximate picture and require no maze of detail. Exponential growths that become saturated and thereby slowed down to a steady level are very common in nature. We meet them in almost every field of biological growth or epidemiology. The rabbit population in Australia or the colony of fruit flies in a bottle all grow rapidly until some natural upper limit is reached. In nearly all known cases, the approach to the ceiling is rather strikingly symmetrical with the growth from the datum line.[7] The curve of growth is a sigmoid or logistic curve, S-shaped, and even above and below its middle.

The only good historical example known to me illustrates the decline of the European Middle Ages, followed by the beginning of the Renaissance. If one makes a graph of the number of universities founded in Europe, arranged by date, the curve splits into two parts. The first part is a sigmoid curve starting at A.D. 950, growing exponentially at first but falling away rapidly by about 1450, and thereafter approaching a ceiling with equal rapidity. Added to this is a second exponential curve, doubling more rapidly than the first and acting as if it had started with a first member, a new style of university in 1450. The lesson is obvious: the old order began to die on its feet and, in doing so, allowed a quite new, renaissance concept of the university to arise.

It is a property of the symmetrical sigmoid curve that its transition from small values to saturated ones is accomplished during the central portion (halfway between floor and ceiling) in a period of time corresponding to only the middle five or six doubling periods (more exactly, 5.8), in-

7. A collection of such sigmoid graphs showing autocatalytic growth is to be found in Alfred J. Lotka, *Elements of Mathematical Biology* (New York, 1956), Ch. 7, figs. 4–8. In the same work, p. 369, fig. 71 is another sigmoid graph, this one indicative of technological rather than scientific growth, that of the total mileage of American railroads. The osculating tangent (straight line through the mid-point of the S-curve) acts as if it is started *ca.* 1860 and attained saturation (at some 300,000 miles) *ca.* 1920. This, then, is the *effective* span of this aspect of the Industrial Revolution in America. At least this method has the advantage over many historical discussions in suggesting some decent and objective criterion for what constitutes an effective beginning and an effective end to the process. The same criterion distinguishes capricious precursors from true originators.

dependent of the exact size of the ceiling involved. Thus, the time at which the logistic curve has fallen only a few per cent below the expected, normal exponential curve represents the onset of the process. Three doubling periods later, the deficiency is 50 per cent, the sigmoid curve reaching only half the expected height. Thereafter, the sigmoid curve becomes almost flat, while the exponential curve continues its wild increase. One must therefore say that only some three doubling periods intervene between the onset of saturation and absolute decrepitude.

For science in the United States, the accurate growth figures show that only about thirty years must elapse between the period when some few per cent of difficulty is felt and

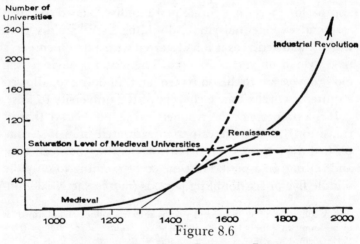

Figure 8.6

NUMBER OF UNIVERSITIES FOUNDED IN EUROPE

From the foundation at Cairo in 950 up to ca. 1460 there is pure exponential growth, doubling in about one hundred years. Thereafter saturation sets in, so that the mid-region of the sigmoid extends from 1300 to ca. 1610. Between 1460 and 1610 is a period of transition to the new form of universities, a growth that also proceeds exponentially as if it had started from unity ca. 1450 and doubling every sixty-six years. There is probably an ever greater transition to yet faster growth starting at the end of the Industrial Revolution.

the time when that trouble has become so acute that it cannot possibly be satisfied.[8] It seems quite apparent from the way in which we have talked, from time to time in recent years, about manpower difficulties in science that we are currently in a period in which the onset of a manpower shortage is beginning to be felt. We are already, roughly speaking, about halfway up the manpower ceiling.

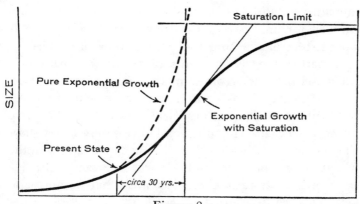

Figure 8.7

The historical evidence leads one to believe that this is no incidental headache that can be cured separately by giving science an aspirin. It is just one symptom of a particularly deep-rooted disease of science. Perhaps it is more a natural process than a disease, though clearly we partici-

8. To be more exact, the standard equation for a sigmoid curve is $y = (1 + e^{-x})^{-1}$ where $x = k \ (t - t_0)$. Its osculating tangent intersects the abscissa at $x = -2$, and the line $y = 1$ at $x = 2$. In the former place, at $x = -2$, $y = 0.12$, whereas the plain exponential function $Y = e^x$ has the value $Y = 0.135$. y is therefore some 12.0 per cent of Y below the value of Y; at $x = 0$ we have $y = 0.5$ but $Y = 1.0$, a falling of some 50 per cent below expectation. Each unit of x corresponds to 1.45 doubling periods, or a factor of $e = 2.718$. The four units of the middle portion of the curve therefore correspond to 5.8 doubling periods, and the interval between a shortage of 12.0 per cent and that of 50 per cent is only 2.9 doubling periods.

pants in the process are ill at ease as a result. It is essential to the nature of the case that science go through a period of vigorous growth and that there has now come a sort of post-adolescent hiatus, and the growth is done and science has its adult stature. We must not expect such growth to continue, and we must not waste time and energy in seeking too many palliatives for an incurable process. In particular, it cannot be worth while sacrificing all else that humanity holds dear in order to allow science to grow unchecked for only one or two more doubling periods. It would seem much more useful to employ our efforts in anticipating the requirements of the new situation in which science has become, in some way, a saturated activity of mankind, taking as high a proportion of our expenditure in brains and money as it can attain. We have not reached that stage quite yet, but it is only a very short time before we will—less than a human generation. In the meantime, we must certainly do what we can to provide the aspirin of more and better scientists, but we must also face the larger issue ahead.

What makes it particularly exciting is that the bending of the curve toward a ceiling is happening just at that time when the handicap race of the various Industrial Revolutions has been run out and ended in a close finish. In previous decades the runners have been far apart; now they are bunched together and their speeds no longer have much effect. To think out the consequences of this, we must now examine the feeling of living in a state of saturated science.

Some of the effects are already apparent and may be amenable to historical analysis and even statistical treatment. If the cumulative expansion of science rapidly outpaces all efforts we can make to feed it with manpower, it means that more and more things will arise naturally in the life of science and require attention that cannot be given. There will be too many discoveries chasing too few workers. At the highest level we must come to a situation at which there are too many breakthroughs per square head. In all previous times, for each breakthrough, such as that of X rays in 1895, there were many large groups of physicists who could attack the new problem and start to work on it. Already in our own times we have a decrease of this. In any particular area of breakthrough there are initially fewer capable specialists, and many of these are faced with the prospect of having too many interesting tidbits on their own plate to feel the need to go elsewhere, however exciting that might seem.

It may be remarked that this specialization may also be measured, and if you do it in any reasonable way, it appears to lead to the result that it, too, is doubling in every decade or so. As the amount of knowledge increases, each man must occupy a smaller and smaller segment of the research front. This, again, is not a process than can continue indefinitely; eventually a point of no return is reached at which the various disadvantages of acute specialization became too marked. Cross-fertilization of fields decreases and so thereby does the utility of the science. The more rapidly moving research front tends to leave behind such specialists, in increasing numbers, to while out their years of decline in occluded pockets.

Thus far nothing has been said about the quality of research as opposed to its quantity. This is, of course, much more difficult to determine and would repay much more serious investigation than it has ever had. Various measures are possible. One may study the growth of only important discoveries, inventions, and scientific laws, rather than all such things, important and trivial; any count of this sort immediately shows that the growth, though still exponential, possesses a doubling time that is much longer than that of the gross growth of science. The actual stature of science, in terms of its achievements, appears to double within about one generation (some thirty years) rather than in the ten years that doubles numbers of papers and numbers of scientists.[9]

In its stature, science grows much more nearly in keeping with all else in our society: size of population, economic wealth, activity in the arts. In size, however, it must undergo something like three doublings for each of these other generations. Perhaps it is not entirely wrong to see this as a consequence of the cumulative structure of science. If it grows like a pile of stones or bricks, then the pile keeps the same pyramidal shape. Its height measures the stature of science and its attainment; in this it grows at the same general rate as our culture at large. However, to make the pyramid twice as high, its volume must be

9. It is difficult to be precise about this law; so far, I feel, one may have only reasonable certainty that the stature of science, however one defines it, grows some two or three times more slowly than any measure of gross size. One need not argue about the exact size of the constant involved. What is particularly impressive is that the cost of science, in terms of expenditure in money and national income, grows much *faster* than the gross size. Indeed, Strong and Benfey suggest (*Journal of Chemical Education, 37,* 1960, p. 29) that United States research and development costs double every six years, whereas the persons listed in *American Men of Science* double only in twelve years. Thus it would seem that the cost goes up as the square of the number of men working, and the number of men increases as the square or cube of their effectiveness in increasing the stature of science. We have therefore a fourth or sixth power law of rapidly diminishing returns. To proceed with rocketry at ten times the present effectiveness would cost say ten thousand or perhaps a million times as much money! To return to the measurement of the stature of science, it may be noted that on the basis of such subjective lists of "important" discoveries as those of L. Darmstaedter and of P. Sorokin, the evidence seems to agree that there was quite normal exponential growth, doubling in about 120 years for all the period up to about 1660, and then again normal growth doubling every thirty years from that time to the present day.

multiplied by eight, the cube of two. It must undergo three doublings for every doubling of the height. The number of bricks of scientific knowledge increases as the cube of the reach of that knowledge.[10]

Even if this is only a most approximate law, based on rather tenuous hypotheses and measurements, it nevertheless constitutes a powerful law of diminishing returns in the world of science. This finding may be easily strengthened by an analysis of the distribution in quality of scientific men. It has been proposed, on the basis of statistical investigations of the number of times that various papers were used by other people, that an inverse square law of goodness holds here as it did for productivity. For every single paper of the first order of importance there are four of secondary quality, nine of the third class, and so on. Much of the same result is obtained if one regards the spread in the scientific population as similar to that as the upper tail of a normal distribution curve of some sort of intelligence quotient.

However you do it, it seems inevitable that to increase the general number of scientists you must cut off a larger section of the tail, rather than increase the thickness of the same section of tail. Probably it follows that to double the population of workers in the few highest categories, there must be added some eight times their number of lesser individuals. At a certain point it becomes rather futile to worry about improving the standard of the low-grade men, since it is unlikely that one can tamper very much with a distribution curve that seems much the same now as it was in the seventeenth century, much the same

10. This "fortuitous" agreement with a popular and picturesque model leads one to wonder whether some of the other highly descriptive phrases which scientists habitually apply to their tactics can have more than casual usefulness. Is, for example, the geographic simile (fields of work, borderline investigations, difficult territory) a suitable description of the topology of the connectivity of learning?

in America as in Europe or as in Russia. Minor differences in quality of training there might be, but to work on the research front of modern science demands a high minimum of excellence.

Thus science in an age of saturation must begin to look rather different from its accustomed state. I believe it is without question that the occurrence of such a change must produce effects at least as disturbing to our way of life as an economic depression. For one thing, any slackening of the research pace of pure science must be reflected quite rapidly in our advancing technology, and thereby in our economic state.[11] It is difficult to say just what form this effect might take. Clearly there is no direct, one-to-one relationship of pure science to technology. Even if there were declared a sudden moratorium on pure scientific research, or (what is more plausible) an embargo on growth that allowed all such work to continue but without the habitual 6 per cent yearly increase in manpower, there would still be enough of a corpus of knowledge to provide for technological applications for several generations to come. As Robert Oppenheimer has expressed it, "We need new knowledge like we need a hole in the head."

There is, however, a snag in the argument as expressed above, for in the past the expansion of science and of technology have proceeded hand in hand, and it has been only the sorry task of the historian to point out examples where the one or the other has taken the leading role—an

11. For such an analysis of the role of research in economic growth, see Raymond H. Ewell, in *Chemical and Engineering News, 33,* No. 29 (July 18, 1955) pp. 2980–5. Ewell makes a good case for the growth rate of individual industries and the gross national product both being directly proportional to the growth rate of expenditure on research and development. In detail, some 10 per cent increase in the cost of science is needed to produce the national economic growth rate of 3 per cent; that is, the scientific budget seems to increase as the cube of the general economic index.

evaluation in most cases that has been revised back and forth several times each decade. I suspect, because of this intimate relationship, that although technology might be left with a great bulk of pure science waiting to be applied, any decrease in the acceleration of science will prove an unaccustomed barrier to industry, and that the flow of new ideas into industry will in some indeterminate way suffer and drop spectacularly. We are now geared to an improvement of technology at a rate of some 6 to 7 per cent per annum, and a decline in this must affect all our lives. Then, again, if manpower is chronically to be in short supply in the world of science, it will follow that *what* we do is much more important than *how much* we do it.

It follows also that the good scientist will be increasingly in demand and in power, since it must become ever more apparent that it is he who holds the purse strings of civilization in the era we have entered. Indeed, if it were not for the well-established reluctance of scientists to enter the political arena, one might boldly predict that the philosophers are about to become kings—or presidents at least.

In a saturated state of science there will be evident need to decide, either by decree or by default, which jobs shall be done and which shall be left open—remembering always that an ever increasing number of possible breakthroughs must be left unexploited. It is most doubtful whether this can be best done by considering merely the utility to society of the job in itself. In the history of science, it is notorious that practical application has often grown out of purely scientific advance; seldom has pure research arisen from a practical application by any direct means. I would be cautious here, for there are too many violent views in such areas, and the truth is certainly no unmixed extreme. But even so, it would be foolhardy to direct all medical research to work on cancer, or all physicists to work on missiles and atomic power.

If such fields are rich and important at the moment, it is evident that they have not always been so, that they will probably appear in a different light a few decades hence. In this future state, we might perchance depend on fields that are currently being starved through diversion of the funds elsewhere. If at any time in the future we wish to change, even if the demand is great, we might have already committed our resources in such a way that they cannot be converted to the new projects. Not only is science changing more and more rapidly; it is entering a completely new state.

In this new state, our civilization will rise or fall according to the tactics and strategy of our application of our scientific efforts. It is anarchical to decide such issues by merely letting ourselves be ruled by the loudest voices. It may or may not be worth while to support missile research to the hilt, but no man can make such a decision without considering the possibility that this work will ruin the chances of half a dozen other fields for an entire generation. In a condition in which so much of our scientific research is supported by military contract and federal projects, it seems no man's business to consider the possible damage which could come in our new saturated state.

If the supply of research cannot simply be allowed to follow the ephemeral demand, it seems also that we can no longer take the word of the scientists on the job. Their evaluation of the importance of their own research must also be unreliable, for they must support their own needs; even in the most ideal situation they can look only at neighboring parts of the research front, for it is not their own business to see the whole picture. Quite apart from the fact that we have no national scientific policy, it is difficult to see any ground on which such a policy might be based. It is difficult to take advice from either the promoters of special jobs or from the scientists themselves, for their interests might well be opposed, might well be irrelevant, to the needs of the nation as a whole.

The trouble seems to be that it is no man's business to understand the general patterns and reactions of science as the economist understands the business world. Given some knowledge of economics, a national business policy can be formulated, decrees can be promulgated, recessions have some chance of being controlled, the electorate can be educated. I do not know, indeed, whether one might in fact understand the crises of modern science so well as to have the power to do anything about them. I must, however, suggest that the petty illnesses of science—its superabundance of literature, its manpower shortages, its increasing specialization, its tendency to deteriorate in quality—all these things are but symptoms of a general disease. That disease is partly understood by the historian, and might be understood better if it were any man's professional province to do so. Even if we could not control the crisis that is almost upon us, there would at least be some satisfaction in understanding what was hitting us.

POSTSCRIPT

The material covered in this chapter has probably undergone more development and change than any other. It rapidly proved to have a life of its own, so that it grew first into a separate book (*Little Science, Big Science,* [New York: Columbia University Press, 1963]) and then touched off a continuing series of research papers exploring many different quantitative investigations based on the counting of journals, papers, authors, and citations. In no time at all there were bibliographies and conventions devoted to

bibliometrics and to *scientometrics,* and even a meeting of
the invisible college of people studying invisible colleges.
Proceeding partly from a debt of inspiration that I paid in
a Festschrift for the late Professor J. D. Bernal,* and partly
from my long-standing and very friendly collaboration with
Gennady Dobrov of the Ukrainian Academy of Sciences,
Kiev, the term "science of science" achieved an almost explosive popularity. Unfortunately, though it came readily
to the tongue and pleased those who desired objective investigations of the workings of science in society, the term
rapidly became debased by being used in as many different
ways as there were users, and by being taken as a promise
to deliver goods that were by their very nature undeliverable.

The field, whatever it be called, has by now attracted a
reasonable number of competent workers who are cumulat-
ing their researches and no longer inventing the wheel in
each his own way. I take the position that the workings of
science in society show to a surprising degree the mechanistic and determinate qualities of science itself, and for this
reason the quantitative social scientific investigation of science is rather more successful and regular than other social
scientific studies. It seems to me that one may have high
hopes of an objective elucidation of the structure of the
scientific research front, an automatic mapping of the fields
in action, with their breakthroughs and their core researchers all evaluated and automatically signaled by citation
analysis. Furthermore, I feel it will be relatively easy to
link such quantitative data with economic and manpower
data available as the fiscal statistics of the inputs fed by
government and industry to scientific research. What is

---

*"The Science of Science," in *The Science of Science,* eds. M. Goldsmith
and A. L. Mackay (London: Souvenir Press, 1964; published in U.S.A.
as *Society and Science,* New York: Simon and Schuster, 1964), pp.
195–208. Pelican edition, London, 1966; Russian edition, Moscow, 1966.

decreasingly clear is the relation of all this to the political process involved in the choice of technologies by society. For such problems we need much closer interaction between my sort of work and that of the economists and political scientists who have been making great strides in these other directions.

Postscript to footnote 6

It was asking for trouble to set up such a record! The present marathon laureate, to the best of my knowledge, is Theodore Dru Alison Cockerell (1866–1948), Professor of Natural History, University of Colorado. He published a grand total of 3,904 items over a period of 66 years, an average of a little more than a paper each week. He worked in short papers, under a shadow of imminent death. (See William A. Weber, University of Colorado Studies, Series in Bibliography, no. 1, Boulder, 1965.)

CHAPTER 9: EPILOGUE

The Humanities of Science

THE WORD "idiot" had its origin in the Greek *Idiōtēs*, a private person, a layman, a nonprofessional, unqualified by nature or nurture for participating in what was then uppermost in the life of mankind—the experiment of political democracy. This term, now sadly debased, might well be recoined to describe our modern *scientific idiots* —those cultivated men who would avert their eyes from science and recoil from what they would take to be a priestly mumbo-jumbo of incomprehensibility surrounding the new growing-tip of civilization, its sciences, and their associated technologies.

The scientific idiocy of modern culture has now been diagnosed by many distinguished anatomists of our present state of melancholy—by Sir Eric Ashby, Jacques Barzun, Herbert Butterfield, George Sarton, and Sir Charles Snow, to mention but a few.[1] There seems general agreement

1. Sir Eric Ashby, *Technology and the Academics* (London, 1959); Jacques Barzun, *Teacher in America* (New York, 1954); Herbert Butterfield, "The History of Science and the Study of History," *Harvard Library Bulletin, 13* (1959), pp. 329-47; George Sarton, *The History of Science and the New Humanism* (New York, 1956); C. P. Snow, *The Two Cultures and*

that any separation of the sciences from the humanities is
a bad thing. The gap must be bridged, or it must be con-
strued out of existence by considering science as a humanity or the humanities as sciences. Our educational system is failing by producing graduates who might well be
awarded certificates of ignorance, either in the humanities
or in the sciences. Our scientists and our humanists are
both becoming deficient for the urgencies of civilization
and scholarship, because of their lack of knowledge on both
sides of the fence.

In the preceding chapters I have tried by exemplification
and by pleading to show that the midregion between the
humanities and the sciences is worthy of serious scholarly
study, that it is exciting, and that it might be useful. Only
by dint of man-size labor may all the traditional modes of
thought of humanistic scholars and all their armory of
techniques for inquiry be brought to bear upon the subject matter of science. This scholarship, moreover, tells
more about science than any mere scientist can learn by
osmosis in the course of his proper studies, and it must
provide whole sections of history, philosophy, economics,
and sociology of science that now exist as scholarly subjects
only in embryo.

This, then, is my first claim. Here is a worthy subject
of scholarship and research, a field in which all the humanistic techniques can be turned upon all the sciences. As

the Scientific Revolution (New York, 1959). Of these books, that by Sarton,
originally published in 1931, was the trailblazer, far ahead of the spirit
of his time. For more specific evaluations of the history of science as a
foe of scientific idiocy and as an autonomous field of scholarship, see
I. Bernard Cohen and Fletcher G. Watson, eds., *General Education in
Science* (Cambridge, Mass., 1952). A more recent set of evaluations and
discussions was presented at a conference on the history, philosophy, and
sociology of science, sponsored by the American Philosophical Society and
the National Science Foundation during February 1955, and published
in the *Proceedings of the American Philosophical Society*, 99 (1955),
327–54.

such, it is the prime duty of any toiler in this field, as in any other, to pursue his studies, publish his monographs, and little by little reproduce his kind by training research students and giving them a guiding light a little brighter than the one that lit his own steps.

One could stop here. The subject would then be accorded full rights as a scholarly autonomy, like any of the other exotic specialties (such as Assyriology, Dante studies, or Geochronology) that are allowed a place in a few of the world's great universities—perhaps even a little institute all to itself. Many would argue that this is the only rational strategy of scholarship. Only those who must study this subject would then find it and (what is perhaps more crucial) contrive some cunning device of foundation grant or peripheral bread-and-butter teaching post that would give them the academic leisure to pursue this devious end. Even more scientists (and technologists and physicians) would wait for their retirement and devote their terminal leisure to being self-made historians, showing all the disadvantages of unskilled labor, making *ex cathedra* statements *about* science—but nevertheless producing, along with the chaff, some grain of first-rate works of high scholarship.

This, in general, is the very way in which history of science and history of medicine (and to some degree philosophy of science) have operated until quite recent times. Clearly there shall always exist this sort of learning while there yet survives honorable place for the lone scholar, for the inspired amateur, and for the retired professional of gentle tastes. The great pioneers in our field were all such men, and my foremost concern is to honor their names, uphold their ideals, and further their teachings. The world of scholarship is not, however, composed exclusively of such men. The universities, colleges, and schools have a social contract by which they engage also in education of the population at large, in its training for lives other than that of single-minded learning, for jobs outside the world of the university. All the great lines of specialization in the humanities and the sciences are taught now to many more students than those that have an urge for this alone. Seen in this light, the academic machine for producing physicists, or historians, or philosophers, or what you will, has a waste product of more than 90 per cent who do not become professionals at the research fronts of knowledge. Our society allows this because we have remarkably good use for this "waste product" in other directions and also because it provides a very good sieve for picking out the bright and productive 10 per cent (or less) in each field.

It seems evident that we need the facility of this big machine for the humanistic examination of science rather than the little machine, minutely efficient though it be, of the Assyriology stage. There is ample precedent for such necessary growth from isolated scholars of esoteric fields into the complex of a large-scale subject, accepted as a normal major department of most sizable colleges; many of our scientific disciplines emerged thus out of the region of natural philosophy. In another direction, the subject of economics might be an excellent parallel. Economics is a particularly apt analogue, for, as we have attempted to show, our discipline tries to do for the scientific world just what economics does for the world of business and commerce.

Only as such a large-scale subject can our discipline act as an educational bridge between the arts and the sciences. Only thus can it produce its own 90 per cent waste product of students who will go out into all those jobs and professions midway between the sciences and the rest of civilization. Only thus can we be sure of attracting, at an early stage, a sufficiency of the first-rate minds of this generation who need some exposure to the humanities of science before they can realize that it is here that they might make their major contribution. Here, then, is my second claim: not only is the subject worthy, but it must be practiced as an autonomous large-scale field of study—not as a rare fragment of specialty.

By insisting that a university department in our discipline must be large, we raise certain difficulties but solve many more. In the first place, only by this device can we increase ourselves beyond the ranks of those few isolated scholars who can acquire special dispensation from foundations and presidents, and those equally few who can claim with enough assurance that they can teach all the range of this subject that the deans and departments seem to require. One is as likely to find a single man to teach humanities of science as a man who can teach all history and all science—less likely, indeed, for in our bailiwick one becomes highly conscious of the contributions of non-Western civilizations and must needs trespass on the lands of the Arabist and the Sinologist. In a reasonably large department one need only insist, and much more possible it becomes, that a man worthy of hiring need have only a good general background plus research-front knowledge of some well-defined area, such as medieval physics, Greek astronomy, seventeenth-century scientific societies, eighteenth-century German medicine, or Lavoisier studies. The same is asked of the graduate student, and at last it all falls within the pattern of normal academic machinery. No longer need the poor migrant to our studies feel it incumbent upon him to write the definitive history of all science, or of some large chunk of it, in order to demonstrate his qualification for calling himself an historian of science. Now, all he should need is good work.

Insisting upon autonomy for the large-scale department creates, however, a special administrative difficulty for universities. Such a department is not born in maturity; it must develop slowly and keep in tune with the traditions and financial possibilities of the institution concerned. At many colleges this has led to the growing-up of such studies within an already flourishing department of history or of philosophy, or from all or one of the science departments. In a few cases it has been successful, and the man appointed has been sufficient of a giant to become recognized as an ornament of scholarship within the larger matrix, a man capable of attracting good students around him and pro-ducing work that meets with approval. In less fortunate cases, the subject becomes recognized only as a minor specialty within history or philosophy, or gets tacked on and hangs precariously to the coattails of the scientists. I do not know which is the lesser of these evils, for when the man is successful, his subject at that institution becomes a one-man show, and his students are often immediately recognizable, not as true scholars in the old man's tradition, but as little facsimiles sharing the master's foibles and enthusiasms. In a field so wide and so ramified as the Humanities of Science [2] we can no longer afford to exist solely in one-man shows. No one man can cover enough of the field with firsthand experience and teach it in sufficient depth to give a fair start to the next generation.

2. It is not within my conscience to apologize for this term, vague and ill-defined as it may seem. It was first used tentatively in a paper read before the second American Humanities Seminar, at Amherst, Mass., in June 1957, later printed as "The Scientific Humanities—an Urgent Program" in *Basic College Quarterly*, Michigan State University (Winter, 1959); and in *The Graduate Journal*, University of Texas (Fall, 1959). It was coined to describe a discipline or academic subject rather than the movement or trend implied by Sarton's "Scientific Humanism." Unfortunately the term Scientology has already been bespoken by the sect of Mr. Ron Hubbard, and Scientistology is too grotesque. Humanities of Science does not come too readily to the lips, but it seems to give the right impression, and I find it far superior to The History, Philosophy, and Sociology etc., of Science, Technology, and Medicine etc.

It follows, therefore, that however convenient it might be for an institution to start the seedling department within the shade of an older, fruit-bearing tree, be it of history, or science or philosophy, this is not calculated to induce vigor. It is better for our subject to stand on its own, contriving and needing good will from all its colleagues. Whether, lacking possibility of direct access, the appointee has approached his subject from the side of the sciences or of the humanities, he must not seek the allegiance of his erstwhile colleagues at the expense of those of the other side. He must strike a middle course, steer by the light of his own discipline, and have faith in its ample integrity. In but one more generation of students we may perchance have enough of those who have grown up within this field primarily and cannot be regarded either as fragmented historians or perverted scientists. For the present we must accept the hazards of our birth. The autonomy of the department is something that can always be insisted upon, but the desirability of having many teachers must bend to the power of the dollar. If only one man can be appointed, let him be good at his trade rather than universal in it. If he knows only about William Harvey, he is probably better than a man who lays claim to the whole of history and philosophy of every science and choice bits of technology and medicine to boot. Only in a world of amateurs could one pretend to such monolithic omniscience.

It is tempting at this point to ask, "Given such a department, autonomous and large-scale, what does one do with it? How does it function?" It is indeed tempting, for, in addition to Yale there are forty-six colleges in the United States where history of science is taught, and (I believe) at twenty-four one can now earn a Ph.D. in the subject or in some combination of it with the philosophy of science. We are all faced with much the same problems, though I must admit that many of my colleagues do not agree with

me about extending the subject to a large scale. Perhaps
they have had too gruesome experiences with the massive
required courses that some universities have instituted to
build the famous educational bridge.

At the level of the graduate school the answer about
methods and aims is difficult but not impossible. Clearly,
the student must come to grips with all or nearly all the
traditional avenues of inquiry in our field,[3] and in doing
this he must learn its special and peculiar techniques as well
as those of the adjacent scientific and humanistic areas.
One cannot demand the impossible—that students should
all become adept in Greek, Latin, Arabic, and Chinese—
but one can reasonably hope to secure a convert from time
to time from other departments with such skills. The same
holds for special scientific or historical skills. A territory
such as ours holds many attractions, and we may hope to
get suitable people.

Two questions seem to need comment with respect to
graduate work: what sort of students does one admit, and
how should their portions of study be allocated? The customary answer to the first point is that the student must
have ample scientific training as a basis, and as much historical feeling as possible; at least this must be the normal
answer until such time as we can produce undergraduates

3. For a summary of the traditional avenues of inquiry in the history
of science one may go either to some good, general, and comprehensive
history of science—*e.g.*, Charles Singer, *A Short History of Scientific Ideas
to 1900* (Oxford, 1959)—or to the several selective bibliographies and
reading lists in the field. Among these I would recommend the following:
Marie Boas, *History of Science*, Publication No. 13 of the Service Center
for Teachers of History, (Washington, 1959); Henry Guerlac, *Science in
Western Civilization, a Syllabus* (New York, 1952); *The Early History of
Science, A Short Handlist* (Helps for Students of History No. 52), The
Historical Association (London, 1950); George Sarton, *Horus, A Guide to
the History of Science* (Waltham, Mass., 1952). The introductory essays in
this last, especially Ch. 3, "Is it Possible to Teach the History of Science?"
should be particularly valuable to those setting about the task of finding
out if it is possible to learn the history of science.

trained in this area from the egg. It is not by any means to be taken as an unexceptionable rule, however, for there exist pathological examples to the contrary. It so happens that three or four of the most brilliant contributors to our studies have entered from the side of the humanities and have demonstrated their clear abilities to absorb and digest the science with an adequacy that is startling. Perhaps it is improbable but it is not impossible, and one must therefore allow for the man who has always shown preference for history, let us say, but has managed to acquire *passim* enough scientific backbone to read the *Scientific American*. Humanists who are worth their salt will attract students other than those who hate science, abhor mathematics, and make themselves scientific idiots.

At the undergraduate level the nature of courses, at this stage of development, is almost certainly experimental. From the point of view of traditional scholarship, they should be oriented so as to be a feeder for the graduate school. A student should be able, given sufficient ability and desire, to pass from his baccalaureate into graduate work in the same or an equivalent department, without needing to take extra years for more science or more history. From the educational standpoint, that of helping to rid ourselves of scientific idiots, it is desirable that the new subject, humanities of science, should provide a matrix that will inject a sufficiency of science into the best possible liberal education as conceived within the framework of the humanities. Better still if the new subject can be the mortar that holds together one part of the humanities and an equal amount of the sciences themselves. If this can be achieved (and I see every prospect of its doing so) it might well provide a more honestly scholarly way of teaching science to "non-specialists" than some previous attempts at General Education in Science. These attempts were very worthy and went part way to a solution, but they seemed to lack

some element, and this lack made them suspect. Perhaps
the new brand of subject matter, picked away from its
tissue of science and history, might provide that element.
To this end, my own proposal would be for a new undergraduate major, composed of about one part of the sciences,
one of the humanities, and one of the history and philosophy of science.

At both the graduate and undergraduate levels there is
need to tackle several questions that I have striven to leave
unresolved by calling the subject "Humanities of Science."
What is the proper allocation and balance between history
and philosophy of science (or scientific method, as it is
sometimes called)? What between the pure sciences and
the technologies? What about the sociology and psychology
of scientists? What of the history of medicine? I would
claim that these parts form an indissoluble complex, most
vexing to dissect. I do not see how anyone can teach history
of science without that of technology and of medicine, and
vice versa. How could one teach the history of Connecticut
without that of the United States, of Europe and the world?

My own personal solution is to have the general histories
of the physical sciences and of the biological sciences as a
basic diet, followed by a selection of excursions into all the
other areas—technology, medicine, American science, medieval science, Islam and the Orient, etc.—as opportunity
and need dictate. My own preference, further, is for a staple
food that has some three parts or four of history to but one
of philosophy, with a rare spicing of sociology and psychology of science. This is based not on any evaluation of
the importance or interest of those respective fields but
merely upon the variety of subject matter and source materials with which the student must familiarize himself.
One can, of course, spend a life's work in but one corner
of any of those sections, but for a good over-all training
the student should have the privilege of being exposed to

as much as possible of all that the world has to offer. It is the advantage of a large-scale department that this can be done more efficiently there than in a one-man show; it is the sweetness of autonomy that the graduating scholar is then qualified in his own right and not as a mere sub-specialist, imperfect as an historian, unproductive as a scientist.

What, you may ask, are we to do with those who come out qualified as Humanists of Science? There is, I believe, an ample choice of answers for this. First, ours is one of the most rapidly growing scholarly disciplines in the United States, perhaps in the world. At each international congress and annual meeting the brotherhood is struck by the increasing number of new converts, a high proportion of them holding posts created since last we met. We shall need, for university teaching posts, many times over the present flock of doctorate graduates from the major institutions producing them. Eventually, too, we shall need high school teachers and teachers of such teachers—for it seems likely that Humanities of Science must to some extent displace science itself at this level as well. At another level, for both graduate and undergraduate students there is an increasing need for administrators of scientific organizations. The learned societies, the national and private foundations, the posts of political responsibility in science, the science attachés at embassies, are all increasing rapidly and assuming a complexity and expertise that is beginning to put them out of range of the plain scientist. Even if they were sufficient for the task and ideally trained for it, we do not have and cannot spare enough scientists to be kicked upstairs from the laboratory bench to the conference table. In industry, as I have been told repeatedly by the large scientific research establishments, the biggest manpower shortage is not at the research front but in the region between there and the front office. Where else can industry get people educated in the best of the liberal tradition but able to talk the language of the scientists and perhaps appreciate more deeply than they do the inner mechanics of their art?

Lastly, as I have tried to show throughout this book, science is part of the central core of our world, and it is a core that is in process of violent change, creaking and grumbling in the process and threatening us with uncontrollable deluges and eruptions. In this age we need an informed and intelligent public to whom science and its workings, even in crisis, is not a mystery. Humanists of science at their research fronts might be able to diagnose the processes, piece together parts of the mechanism of science, but only a public exposed in the colleges and schools to their findings about science can appreciate the depth and import of this cumulative activity that sets our culture apart from all that has come before.

Index